T0259093

EVALUATING ENVIRONMENTAL AND SOCIAL IMPACT ASSESSMENT IN DEVELOPING COUNTRIES

EVALUATING ENVIRONMENTAL AND SOCIAL IMPACT ASSESSMENT IN DEVELOPING COUNTRIES

SALIM MOMTAZ

S M ZOBAIDUL KABIR

AMSTERDAM • BOSTON • HEIDELBERG • LONDON • NEW YORK • OXFORD
PARIS • SAN DIEGO • SAN FRANCISCO • SYDNEY • TOKYO

ELSEVIER

Elsevier
225, Wyman Street, Waltham, MA 02451, USA
The Boulevard, Langford Lane, Kidlington, Oxford OX5 1 GB, UK
Radarweg 29, PO Box 211, 1000 AE Amsterdam, The Netherlands

Library of Congress Cataloging-in-Publication Data
Application Submitted

British Library Cataloguing in Publication Data
A catalogue record for this book is available from the British Library

ISBN: 978-0-12-408129-1

For information on all Elsevier publications
visit our web site at store.eslevier.com

Working together
to grow libraries in
developing countries

www.elsevier.com • www.bookaid.org

CONTENTS

LIST OF TABLES

LIST OF FIGURES

LIST OF MAPS

LIST OF PHOTOS

LIST OF ABBREVIATIONS

ADB	Asian Development Bank
BDT	Bangladesh Taka
BWDB	Bangladesh Water Development Board
CEGIS	Centre on Environmental and Geographic Information System
DG	Director General
DOE	Department of Environment
DOF	Department of Forest
ECA	Environmental Conservation Act
ECC	Environmental Clearance Certificate
ECR	Environmental Conservation Rules
EIA	Environmental Impact Assessment
EIS	Environmental Impact Statement
EMP	Environmental Management Plan
FAP	Flood Action Plan
GOB	Government of Bangladesh
IAIA	International Association of Impact Assessment
IEE	Initial Environmental examination
KJDRP	Khulna-Jessore Drainage Rehabilitation Project
LGED	Local Government and Engineering Department
MOEF	Ministry of Environment and Forest
NEPA	National Environmental Policy Act
NGOs	Nongovernmental Organizations
SEA	Strategic Environmental Assessment
SIA	Social Impact Assessment
TBM	Tidal Basin Management
TOR	Terms of Reference
TRM	Tidal River Management

LIST OF APPENDICES

PREFACE

This book is the culmination of the authors' research and fieldworks on the concepts, evolution, and applications of environmental and social impact assessment (E&SIA). E&SIA is now a well-established planning instrument, both in developed and in developing worlds, intended to minimize and manage impacts of development actions on the environment and society. This environmental management tool has come a long way since its inception through National Environmental Policy Act of 1969 in the United States of America. During this period, authors have attempted to measure its effectiveness in different countries and jurisdictions using various methods. Although the emphasis has been on the examination of one particular aspect (i.e., legal status, quality of environmental impact statements, etc.) of EIA and SIA, there has been growing concern about the absence of a framework that allows examination of all aspects of EIA, from legal status, to actually conducting EIA and preparing EIS, and finally to evaluate the effectiveness of mitigation measures at the post-EIS level. This book fills this gap. Not only does this book provide an integrated holistic framework for evaluating EIA and SIA in developing countries, but it also demonstrates the framework's application using empirical evidences from Bangladesh. This book is intended for academics, researchers, practitioners, and students in E&SIA, environmental science and management, development studies, and political economy. International development agencies and nongovernmental organizations will also find this book a very useful manual to assist in their development impact assessments.

We would like to express our gratitude to the numerous individuals, villagers, officials, academics, and researchers who participated in our field investigations and provided valuable information. Our thanks go to the organizations that provided supports for this study. Some names are listed here: Bangladesh Centre for Advanced Studies (BCAS), World Bank and Asian Development Bank Bangladesh country offices, United Nations Development Program (UNDP) country office, and Centre for Environmental and Geographic Information Services (CEGIS) Dhaka office. The book has been written under the auspices of the University of Newcastle, Australia. We would like to thank our employer for providing us with all supports we needed during the writing of this book. Finally, our deepest gratitude goes to our families for being there for us and supporting us in this important task.

Salim Momtaz
Senior Lecturer, School of Environmental & Life Sciences,
University of Newcastle, Australia

S.M. Zobaidul Kabir
Post-Doctoral Research Fellow, Centre for Environmental
Management, CQUniversity, Australia

AUTHOR BIOGRAPHY

 **Dr. Salim Momtaz** is a senior lecturer at the University of Newcastle, Australia. He teaches in the area of Sustainable Resource Management. He received his BSc and MSc degrees in Geography from the University of Dhaka, Bangladesh. He did a PhD in Regional Planning and Development from the University of London under a Commonwealth Scholarship, working under the supervision of Professor Richard Munton. His academic career started at the University of Dhaka in 1986. Salim moved to Australia in 1994 as an independent migrant. From 1995 to 1998, Salim taught Geography and Environmental Studies at Central Queensland University. He joined the University of Newcastle in 1999 where he has been teaching since. He had a stint in the US teaching Environmental and Social Impact Assessment at Georgetown University, Washington, DC, as a visiting professor. He received Rotary International Ambassadorial Fellowship to teach and conduct research in Bangladesh. Salim's current research interests include development and environment, climate change adaptation, environmental governance, and social impact assessment. Salim led the team that conducted one of the first social impact assessment studies in Australia titled *Independent Social Impact Assessment: Proposed Castle Hope Dam and Awoonga Dam, Queensland*. Salim published five books and many articles in international journals. He was a member of the Scientific Advisory Committee, Netherlands Government Research Organization, between 2007 and 2010. Salim currently lives in a coastal outer suburb of Sydney, Australia, with his wife and two daughters.

 **Dr. S.M. Zobaidul Kabir** is currently working at the Centre for Environmental Management (CEM), CQUniversity, Australia, as a Postdoctoral Research Fellow in the area of environmental and social impact assessment. In addition to research, he teaches social and economic impact assessment at CQUniversity under Environmental Management program. Recently, he has obtained Doctor of Philosophy on Sustainable Resource Management from the University of Newcastle, Australia, under the prestigious Endeavour Postgraduate Award. Earlier, he obtained Master of Environmental Management and Master of Diplomacy from the Australian National University under AusAID scholarship.

Dr. Kabir has more than 7 years of research experience in the area of ex-post evaluation of environmental performance of development projects, environmental policy analysis, environmental and social impact assessment of development projects, and social appraisal of mine closure plan. He worked for the Government of Bangladesh as a civil servant and held various positions. He also worked for the United Nations Population Fund (UNFPA), Bangladesh, as a National Project Professional Personnel (NPPP).

He has more than 12 publications as scholarly journal articles and peer-reviewed conference proceedings. His research interests include impact assessment, greening business, community engagement, liveability assessment, environmental governance and sustainability assessment. He is a member of International Association of Impact Assessment (IAIA) and Environmental Institute of Australia and New Zealand.

CHAPTER 1

Introduction

Environmental Impact Assessment (EIA) has its origin in the passage of National Environmental Policy Act of 1969 (NEPA '69) in the USA. It emerged from the realization that many projects funded by the government in the United States failed to take their environmental impacts into consideration in the development and implementation phase and, as a result, caused major environmental problems. Cost-benefit analysis and other environmental safeguards in place at the time were not adequate for the protection of the environment. This awakening of public conscience regarding environment was happening in the backdrop of the publication of Rachael Curson's book titled *Silent Spring* (1962), the initial activities of the Club of Rome (Meadows et al., 1972), and the first wave of environmental movement in the United States of America throughout the 1960s. NEPA '69 took effect from January 1970 and provided the basis for similar legislations around the world in the 1970s and 1980s. Developed countries (examples include Canada, European countries, Australia, New Zealand) were quick to accept the concept of EIA and provided legal mandate and/or administrative ruling in favor of EIA. For the developing world, the Asian countries like Indonesia, Taiwan, The Philippines, Singapore, and Hong Kong were at the forefront of EIA adoption. Eventually, other Asian countries joined the club of EIA in the 1990s. African and Latin American countries followed suit. Today, most countries on earth have some form of EIA in place. EIA as a preventative environmental management tool has now been well recognized by the governments of the developed and the developing nations and is well embedded in the planning process.

Bangladesh is a latecomer in the EIA arena at least in a formal way. Some form of EIA was in place and used in 1985 in the planning of the Jamuna River Multipurpose Bridge project (the biggest bridge construction in Bangladesh) where the World Bank and the Asian Development Bank as the funding organizations helped conduct EIA of the proposed project. There were also some environmental laws (Pollution Control Act) for environmental protection. In 1992, the first EIA guidelines for the water sector were published. However, it was not until 1995 that the country passed its first EIA legislation titled *Environmental Conservation Act (ECA '95)* (DOE, 1995) to be followed by

Evaluating Environmental and Social Impact Assessment in Developing Countries
http://dx.doi.org/10.1016/B978-0-12-408129-1.00001-2

Environmental Conservation Rules 1997 (ECR '97) (DOE, 1997). For the first time in the history of Bangladesh, ECA '95 legally required all development projects to systematically consider in advance their environmental consequences. The Rules (ECR '97) further clarified the provisions of the Act and were intended to facilitate the enforcement of ECR '95.

The objective of this book is to evaluate the effectiveness of EIA system in Bangladesh using a new comprehensive approach termed by the authors as "integrated holistic framework." The holistic approach adopted in the analysis of EIA system in this book looks into all aspects of the EIA procedure including legislative and administrative status; implementation of EIA, that is, conduct of EIA and SIA (social impact assessment); the quality of environmental and social impact statements (EISs) and public participation; and post-EIS follow-up, that is, implementation of mitigation and monitoring measures. This evaluation study is different to the reviews conducted elsewhere in developed and developing countries that basically looked into one aspect (EIA legislation and administration) or the other (EIS quality) of an EIA system to determine its effectiveness. EIA follow-up is still the weakest link in the EIA system, especially in developing countries. A new approach is particularly important for developing countries as they have now gained considerable experiences in EIA practice which is now due for a major review.

Chapter 2 provides a thorough review of existing approaches in EIA review to emphasize the importance of a new holistic approach. The aim of this chapter is to propose a holistic framework that can be applied to understanding the effectiveness of EIA system in developing countries. Current literature shows that the studies of EIA effectiveness often tend to consider a single stage that can only reveal partial view of success or failure of EIA rather than the whole picture. Based on the review of important aspects of EIA effectiveness under a number of major empirical models, this chapter develops an integrated holistic framework to measure the effectiveness of EIA system in Bangladesh. It can be claimed that putting the models together and capitalizing on their strengths can lead to a complete and stronger framework that facilitates a systematic investigation of EIA effectiveness. It is envisaged that the application of the framework can provide a better understanding of the effectiveness of EIA and promote more effective EIA system in developing countries.

Chapter 3 examines the origin and evolution of EIA in Bangladesh and critiques its present status. Environmental Conservation Act 1995 and Environmental Conservation Rules 1997 have been thoroughly examined. A thorough examination of institutional framework is done in order to identify its weaknesses and strengths.

Chapter 4 thoroughly examines the quality of environmental impact statements (EISs) in Bangladesh using a review package. Review packages have been widely used to examine the quality of EISs prepared in different countries. The authors of this book have used the package developed by Lee and Colley (1992) with significant modifications to suit the Bangladesh context—a process that may also help the developing countries to formulate their own review package.

Chapter 5 examines the emergence of SIA within the framework of EIA. In this chapter, legal and institutional aspects of SIA have been examined and the role of various government and nongovernmental organizations including donor agencies has been discussed. Evaluation of SIA system of a country is relatively new and has started to emerge in the literature lately. Using a modified version of a newly developed review package, this chapter evaluates the SIA system in Bangladesh and reviews the quality of social impact statements. The approach employed to understand the SIA system in Bangladesh can also be adopted in developing countries with similar socioeconomic contexts where SIA is emerging as an important social sustainability safeguard.

Chapter 6 examines the legal and administrative status of community consultation in environmental decision making, especially in the conduct of EIA and SIA. It examines how various environmental organizations—government and nongovernmental—are involving community in their EIAs and SIAs and what roles the donor agencies are playing in its implementation. The notion of public participation and its application in environmental management in Bangladesh is different to that of developed world. This is also true for most developing countries where historically environmental decision making has been a top-down process and the concept of public involvement was nonexistent. The developing countries can learn from the experiences of Bangladesh as described in the pages of this chapter.

Chapter 7 focuses on the last major aspect in the evaluation of the EIA system in Bangladesh—post-EIS or EIA follow-up. Three major projects from three different sectors have been identified for this study. These projects had gone through EIA process, had EISs prepared, and have been in operation for the past few years. This has allowed us to examine whether they have properly implemented mitigation measures, whether they have adequate monitoring mechanisms as suggested by the environmental management plan outlined in the respective EISs, and whether they have followed the community consultation and involvement as outlined in the recommendations of the EISs. This exercise allows us to investigate if the EIA system in Bangladesh is effectively working at the post-EIS level.

Finally, Chapter 8 provides an overview of the book, proposes a framework of an effective EIA system, and concludes the book with recommendations.

Despite the fact that Bangladesh is a latecomer in the EIA arena, it has gained significant experience in EIA since its inception in the 1980s. The country represents ecological and socioeconomic characteristics that are common to many Asian developing countries and some African countries, especially the countries that are emerging as EIA practicing countries. It is believed that the lessons from Bangladesh as learned in the pages of this book can provide useful insights for those countries to launch and expand their environmental management pursuits.

This book is based on the authors' in-depth knowledge of theory and practice of EIA and SIA and extensive fieldwork experiences in Bangladesh in the past several years. The methods employed are questionnaire survey of relevant people; interviews (face to face and with the use of voice recorder) of practitioners in environmental management; and thorough review of EIA and SIA legislations, documents, guidelines, statements; and field visits to a number of ongoing and completed development projects. The methods have been further elaborated in the relevant chapters where necessary. For example, in Chapter 4 (review of EIS quality), an extended explanation has been given about the development of a review package for Bangladesh and how the EISs were physically reviewed using that package.

REFERENCES

Curson R. Silent spring. London: Hamish Hamilton; 1962.
Department of Environment (DOE). The Environmental Conservation Act. Dhaka, Bangladesh: Ministry of Environment and Forest; 1995.
Department of Environment (DOE). The Environmental Conservation Rules. Dhaka, Bangladesh: Ministry of Environment and Forest; 1997.
Lee, N & Colley, R 1992, Reviewing the quality of Environmental Assessments, Occasional paper, Number-24, EIA Centre, University of Manchester, UK.
Meadows DH, Meadows LM, Randers J, Behrems WW. The Limits to Growth: a report to the club of Rome's project on the predicament of mankind. London: Angus & Robertson; 1972.

CHAPTER 2

Evaluating the Effectiveness of Environmental Impact Assessment System in Developing Countries: The Need for an Integrated Holistic Approach

Contents

2.1 INTRODUCTION

This chapter starts with a review of the status of environmental impact assessment (EIA) systems in developing countries. It conceptualizes what constitutes an effective EIA system and examines various approaches to effective EIA systems. An integrated holistic approach is then proposed.

Evaluating Environmental and Social Impact Assessment in Developing Countries © 2013 Elsevier Inc.
http://dx.doi.org/10.1016/B978-0-12-408129-1.00002-4

2.2 STATUS OF EIA SYSTEMS IN DEVELOPING COUNTRIES

The evolution of EIA systems in developing countries differs from that of developed countries (Briffett, 1999; Wood, 2003). The first EIAs to be carried out in developing countries were predominantly in response to the pressure from development assistance agencies on a project-by-project basis. On the other hand, EIA was introduced in developed countries mostly in response to the widespread local demand for better environmental protection (Doberstein, 2003; Wood, 2003). In the later years, the emergence of sustainable development agenda influenced developing countries toward adopting EIA.

A good number of research has been conducted on EIA systems in developing countries (Ahmad and Wood, 2002; Appiah-Opoku, 2001; Briffett, 1999; Doberstein, 2003; Ebisemiju, 1993; Glasson and Salvador, 2000; Kakonge and Imevbore, 1993; Lee and George, 2000; Lim, 1985; Lohani et al., 1997; Nadeem and Hameed, 2008; Paliwal, 2006; Ross, 1994; Sadler, 1996; Tongcumpou and Harvey, 1994; Turnbull, 2003; Wang et al., 2003; Wood, 2003; Zeremariam and Quinn, 2007; Zubair, 2001). These studies show, in general, that the operation and performance of EIA practice in developing countries usually fall behind that of developed countries. Table 2.1 presents a summary of views on the status of EIA practice in developing countries.

Table 2.1 Status of EIA systems and practices in developing countries

Authors	Country	Major deficiencies in EIA systems
Lim (1985)	Philippines, Korea, and Brazil	Actual performance of EIA is significantly diverse from the objectives of EIA; EIA is not fully integrated in planning process; limited authority of review agency
Kakonge and Imevbore (1993)	African countries	Lack of formal legislation for EIA and institutional framework; shortage of manpower and inadequate training; high cost of EIA, inadequate baseline data; and public participation
Ross (1994)	Philippines	Uses EIA as an approval tool; political interference in EIA review; corruption of environmental agency; proponents' negative attitude; and limited role of donor agencies
Tongcumpou and Harvey (1994)	Thailand	Institutional weaknesses; limited judicial and public control; weak position of environmental agency

Table 2.1 Status of EIA systems and practices in developing countries—cont'd

Authors	Country	Major deficiencies in EIA systems
Briffett (1999)	Southeast Asian countries	Lack of administrative and legal arrangements; bureaucratic attitude; inadequate assessment of impacts; community participation
Glasson and Salvador (2000)	Brazil	Inadequate monitoring; bureaucratic approval process; and weak institutional arrangements
Appiah-Opoku (2001)	Ghana	Baseline information; lack of experts; lack of environmental awareness and institutional-review of EIA
Ahmad and Wood (2002)	Egypt, Turkey and Tunisia	Inadequate implementation of mitigation measure; inadequate review of EIA reports and Terms of Reference (TOR); limited public participation; lack of interagency cooperation, sector specific guidelines, and training
Turnbull (2003)	Fiji	Lack of consideration of alternatives; monitoring outcomes; bureaucratic and political culture
Wang et al. (2003)	China	Inadequate analysis of alternatives; limited public involvement; poor impact prediction, mitigation and monitoring; and inadequate institutional infrastructure
Doberstein (2003)	Vietnam	Poor quality of EIS; poor public involvement; limited post-EIA monitoring; inadequate institutional capacity including bureaucratic weaknesses
Wood (2003)	Developing countries	Poor analysis of alternatives; poor EIA reports; limited community participation; inadequate mitigation measures and monitoring
Zubair (2001) and Vidyaratne (2006)	Sri Lanka	Inadequate knowledge of officers; financial constraints; complex project approval process; professional control of consultants; weak infrastructure for enforcing EIA regulations
Paliwal (2006)	India	Poorly defined screening and scoping; insufficient baseline data; inconsistent application of impact assessment methods; improper monitoring and implementation; limited public participation; poor EIA reports; and lack of coordination
Zeremariam and Quinn (2007)	Eretria	Lack of EIA guidelines, limited monitoring, and poor quality of EISs (EIA reports)

Compiled by the authors from the sources listed in this table.

The studies in Table 2.1 show that the EIA system in developing countries is generally weak and the extent of progress of EIA practice differs from country to country. In general, the common issues with EIA in developing countries are (1) inadequate institutional arrangements (legal, administrative, and procedural); (2) poor quality of information in the EIA reports (deficiencies in the implementation stages of the EIA process); and (3) poor implementation of mitigation measures. In addition, further deficiencies include broader contextual factors such as political will, bureaucratic culture, and environmental awareness among the proponents and local community.

Requirements for an effective EIA system in both developed and developing countries are relatively similar, although the level and extent of problems related to the implementation of EIA are different. A common inference in EIA literature is that, institutionally, EIA is already well established in developed countries; the only challenge is to address the technical shortcomings of EIA. On the other hand, the EIA systems of developing countries are characterized by poor socioeconomic conditions in which they are practiced, lack of organized public pressure, and lack of proenvironmental political situation (Ebisemiju, 1993; Lim, 1985; Wang et al., 2003). The extent and coverage of the above-mentioned shortcomings of EIA systems also vary from jurisdiction to jurisdiction of EIA within developing countries. Some developing countries have successfully overcome some of these deficiencies over time, while other developing countries still experience shortcomings in the EIA system.

2.3 CONCEPTS OF EFFECTIVE EIA SYSTEMS

The concept of effective EIA system or the effectiveness of EIA systems[1] is widely studied in EIA literature (Ortolano et al., 1987; Sadler, 1996). However, there is no set and commonly agreed definition of effective EIA system. The particular definitions adopted by authors depend on the level of investigation, and their perception about the roles and goals of the EIA and the context in which the EIA operates.

Sadler (1996) defines EIA effectiveness as the extent to which an EIA achieves its intended goals. However, the extent of goals achieved by an EIA depends on how well it works. Again, how well an EIA system can work depends on a number of issues that ultimately constitute an effective EIA system. Broadly, these issues involve procedural issues and outcome

---

[1] "Effective EIA system" and "effectiveness of EIA" will be used interchangeably in this book.

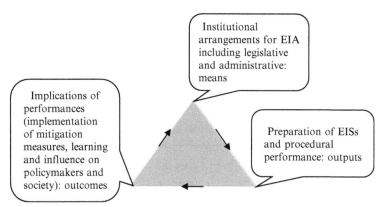

Figure 2.1 Dimensions of an effective EIA system. *Compiled by the authors from field discussions and information; Doyle and Sadler (1996).*

issues in addition to institutional issues. Thus, to understand the effectiveness of an EIA, one needs to look at both "means" that enable the EIA system to work well and "ends" (outcomes) that should be achieved by the EIA. With this in mind, an effective EIA system can simply be defined as one that includes three major dimensions (Figure 2.1): (1) adequate institutional arrangements for EIA, SIA (social impact assessment), and public participation (means); (2) the quality of environmental impact statements (EISs) (outputs of EIA process including social impacts); and (3) implementation of mitigation measures and improvement of natural, environmental, and social well-being (outcomes).

The relevant literature describes the elements or criteria that can define an effective EIA system (Briffett, 1999; Doyle and Sadler, 1996; Ortolano et al., 1987). The elements of effective EIA system as stated by individual authors, however, are not always exhaustive and are often overlapping. In general, literature shows that authors tend to conceptualize the elements that are required for an effective EIA system and for when the EIA system becomes effective.

2.4 DEVELOPMENT OF A CONCEPTUAL FRAMEWORK TO EVALUATE EIA SYSTEM

A significant body of literature has developed on approaches and ways to understand and evaluate an effective EIA system. Four major approaches have been examined below.

2.4.1 Institutional Control (Administrative and Legislative) Approach

The studies by Abracosa and Ortolano (1987), Briffett (1999), Brown et al. (1991), Ebisemiju (1993), and Ortolano et al. (1987) outline the institutional aspects of EIA effectiveness. This approach proposes that an EIA affects the way administrative institutions and legislative arrangements behave. According to this approach, the authors argue that an EIA works well when there is a strong control mechanism including solid administrative and legislative support. The bottom line of this mechanism is to control the proponents' activity within the EIA process. Under judicial control, for example, proponents tend to prepare better EIA reports and comply more with EIA procedure as they are in fear of public litigation. Ortolano et al. (1987) have proposed six ways of controlling the EIA process (Table 2.2) and assert that the presence of this control mechanism may assure the quality of an EIA.

One important aspect of the above approach is that it proposes a set of command and control arrangements including clear and mandatory legal requirements. This clear legislation and ethical code of conduct may force the proponents to comply properly with the EIA requirements (Ebisemiju, 1993) and thereby meet the expectations of other stakeholders. On the other hand, the absence of clear legal directions may lead to general

Table 2.2 Control mechanisms for an effective EIA system

Procedural control: Centralized environmental administrative unit promulgates the EIA requirements, *but does not have the power to modify the project*

Judicial control: Court has power to judge allegations of inadequate attention to the EIA, *but does not have direct control over the project proponents in relations to EIA compliance*

Evaluation control: Centralized administrative unit issues recommendations to decision makers based on an appraisal of the proposed project and the EIA, *but proponents may not comply with the conditions of approval when they are powerful*

Professional control: Project planners have professional standards and codes of ethical behavior that lead them to undertake EIAs for proposed project, *but may not have control over EIA consultants*

Instrumental control: Multilateral or bilateral lending institution requires an EIA before it makes a final decision to release fund for a project, *but both proponents and EIA consultants are not under its direct control*

Direct public and agency control: Citizens, nongovernmental or government agencies apply pressure to influence the EIA process, *but outside the context of above listed controls*

Source: Developed by the authors after Ortolano et al. (1987).

avoidance of requirements, especially where fines and penalties are minimal (Briffett, 1999). These mandatory legal arrangements, including the statutory principles of EIA, are necessary for managing the EIA process effectively, particularly in the context of developing countries.

This approach has some shortcomings. First, this model is restricted to the administrative procedure, ignoring the broader sociopolitical context where the EIA operates. The administration of an EIA may be faced with a variety of contextual forces. These forces may exert influence on the behavior of administrative controls. Therefore, the application of this approach to evaluate an EIA system without considering the context may result in a limited picture of the system.

Second, this approach does not focus explicitly on the power relations among the agencies, an important yard stick against which to assess the effectiveness of an EIA. The proponents may often have more influence than the administrative agency because of the position of the administrative agency in government hierarchy or political preference. In this case, the administrative control may face difficulties in ensuring an effective EIA procedure. This may be called "bargaining position" as reflected in the ability of an agency to exploit its position for the achievement of its mandated goals. The better the bargaining position, the more capable this position is in attaining its goals.

Third, this approach does not consider explicitly the importance of interagency cooperation for effective EIA. Despite having adequate administrative capacity, knowledge, and resources, an EIA process may not be effective in the absence of cooperation and effective networking between proponents, consultants, and administrative agencies. It is assumed that, if there is a more cooperative atmosphere among the actors involved in the EIA process, the EIA system will be more effective.

Finally, this approach constitutes only a partial framework of an effective EIA system. This is because it focuses only on the organizational management aspects of an EIA and pays insufficient attention to technical and scientific aspects, specially the quality of EIS and its contribution in the decision-making process. Furthermore, it does not take into account the implementation of mitigation measures at post-EIS stage, which is an important constituent of an effective EIA.

2.4.2 Decision Making and Quality of EIS Approach

An EIA system is effective when the EISs influence decision makers to take account of environmental issues into project planning. Many studies endorse this viewpoint (Blackmore et al., 1997; Sadler, 1996). The central theme of this approach is that an EIA report should provide decision makers with adequate scientific information so that they can take decisions that are based on complete understanding of project impacts. The assumption here is that a good EIS will lead to better decisions.

This approach has some limitations too. First, it is preoccupied with the idea that a decision can be based only on a good quality EIS.[2] The advocates of this approach are criticized for ignoring the post-EIS stage of EIA (Cashmore et al., 2004). Preparation of an EIS and the consent decision based on it is only half way of an EIA process. Based on the predicted potential in an EIS, decision makers may reject or approve the project with conditions. These conditions may include that the proponents have to modify the project design to avoid or minimize impacts, or adopt appropriate mitigation measures to address environmental impacts. However, modification of a project design and the adoption of mitigation measures in an EIS are not a guarantee that the EIA process as a whole is working effectively unless the mitigation measures are properly implemented at the post–EIS stage of EIA process. This approach thus largely ignores the postapproval (post-EIS stage) stage where the substantive outcomes of an EIA occur in terms of the effective implementation of mitigation measures and protection of the environment (Cashmore et al., 2004).

Second, the decision-making approach assumes that the decision that is based on EIS is rational, as the technical information provided in the EISs are value free. However, instead of a value-free world for information, a scenario of interwoven facts and values is the reality (Owens et al., 2004). Decisions are likely to depend more upon other underlying interests, reflecting the norms and values of decision makers, who usually operate within a political arena. An EIA is a science as well as an art. It is value-laden and ethical in nature (Lawrence, 1997) and, therefore, is placed firmly within the sociopolitical realms of planning that usually seek to make a difference in decision making. Taking decisions on the basis of technical information is not always possible and, therefore, measuring the influence of EIA reports on decision makers may be misleading.

2.4.3 Contextual Approach

Under the decision-making approach, EIA has been assumed to operate within an institutional, structural, and political vacuum (Cashmore et al., 2004).

---

[2] EIS or EIA report will be used interchangeably in this book.

This tendency of trying to isolate contextual issues and favor a more technocratic interpretation of EIA effectiveness has already shown its limitations, as argued by Bina (2007). Contextual variables frequently influence EIA outcomes. Indeed, in several instances, contextual variables are considerably more influential than environmental assessment procedures (Cashmore et al., 2007).

2.4.4 Social System Approach

In a recent theoretical study of EIA effectiveness, Nooteboom (2007) defined the effectiveness of EIA as a phenomenon of large-scale social-learning process. This includes acknowledgment of adaptive behavior such as awareness that cooperation is needed in EIA procedure. According to Nooteboom (2007), system (joint) thinking helps to analyze a situation and to develop an intervention that contributes to sustainable development. Under this model, the EIA is increasingly acknowledged as an instrument for positive feedback, contributing to a growth of cooperation for long-term interest in social capital and government capacity. This approach proposes "sustainable development as dependent on incremental steps that act as levers for breakthroughs to a next level of change or outcomes" (Nooteboom, 2007, p. 48).

No individual approach, as discussed above, is exhaustive in constructing and evaluating an effective EIA system. Each approach helps to explain an effective EIA system, but only reveals a partial picture. Despite some limitations, each of the approaches has individual strengths. A combination of the strengths of these approaches can address their shortcomings and may lead to the development of a strong and complete framework for an effective EIA system. These approaches, however, are not necessarily mutually exclusive. They are complementary and interactive. It is, therefore, imperative to integrate the approaches. This can lead to understanding a comprehensive picture of an effective EIA system.

Empirical studies explicitly using the holistic view in EIA system evaluation are rare in EIA literature. A reductionist view that takes into account one component of an EIA system cannot see the whole problem. It only identifies the problem by reducing it into smaller parts. On the other hand, a holistic view facilitates identification of problems from a broader context and helps to find an appropriate solution. Therefore, we need to move from mere component analysis to more holistic approaches wherein interactive, integrative, and emergent properties are included (Odum, 1977, p. 195).

This holistic view allows researchers to explore complex behavior through a detailed investigation of its components and their interactions. Finally, a holistic view is now mandated if we want to have in-depth insight into the whole EIA systems of developing countries.

2.5 THE PROPOSED FRAMEWORK

In this study, the holistic framework (Figure 2.2) derives from the integration of previously reviewed approaches including institutional aspects of EIA, SIA, and public participation; the quality of EISs; and the consent decision, contextual aspects, and complex social systems. In addition, there is a need to integrate the post-EIS stage of the EIA process in order to have a complete understanding of an EIA system. Thus, the evaluation of the

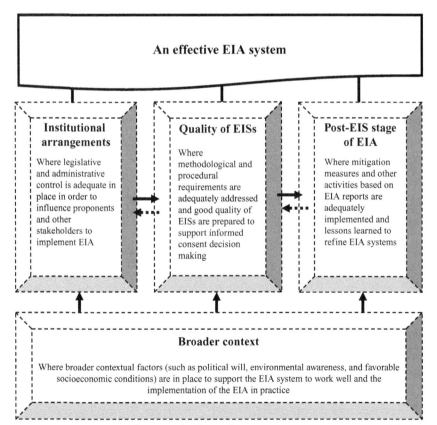

Figure 2.2 A conceptual framework of an effective EIA system: an integrated holistic approach. *Compiled by the authors.*

EIA system using these approaches, together with post-EIS stage of the EIA process, gives us a broader picture. Indeed, putting the essence of these models together under one umbrella provides a framework for broader understanding of an EIA system.

This holistic view of EIA effectiveness recognizes that an adequate institutional arrangement can lead to an effective EIA process that generates a good quality EIA report. The presence of an adequate institutional arrangement and quality EIA report may lead to an adequate implementation of mitigation measures and lessons learned (substantive outcomes). Above all, the broader contextual factors work as a backdrop to strong institutional arrangements, quality EISs, and implementation of activities at the post-EIS stage of the EIA process.

2.6 KEY AREAS AND CRITERIA FOR EVALUATING THE EIA SYSTEM

In line with the above integrated holistic conceptual framework, this study aims to look at three key areas of the EIA system: the institutional framework (arrangements), the quality of EISs, and the post-EIS stage of the EIA process. In the context of Bangladesh, the importance of the evaluation of these three key areas is further illustrated here.

2.6.1 Institutional Framework of the EIA System

The institutional arrangement is one of the fundamental determinants of an effective EIA (Sadler, 1996, p. 78). An effective EIA system along with SIA and the essential component of public participation depend on legal (laws, rules, regulations, and traditions) and administrative (EIA procedures, administrative structures, the roles and responsibilities of agencies, resources, and capacity) arrangements. In fact, the strength of the legal mandate of EIA institutions is a measure of the level of a country's commitment to an effective EIA system (Lohani et al., 1997). It is necessary to scan the current legislative arrangements as an inadequate and ambiguous legislation may impose substantial costs, uncertainty in the application of EIAs, and cause delays in achieving objectives.

In order to be effective, an EIA system needs also to be accompanied by adequate administrative arrangements (Abaza et al., 2004; Ebisemiju, 1993; Wood, 1995). Therefore, a review of the responsibilities, roles, and structures of formal and supportive institutions involved in the EIA process and their capacities for managing EIA practice is also essential. Such a review helps to identify the gaps and strengths of existing institutional arrangements.

Furthermore, the reason behind the study of institutional arrangements in Bangladesh is that the EIA system is relatively new. Therefore, institutional arrangements are likely to be weak as in so many other developing countries. However, there has been no major review of legal and administrative arrangements of the EIA after its formal introduction in Bangladesh in 1995. So, it is not unlikely that the institutional arrangements of EIA system in Bangladesh will have some deficiencies. These limitations need to be explored so that suggestions can be made for adequate institutional arrangements in order that the EIA be successfully implemented.

2.6.2 The Quality of EIS

An EIA report is the most important and tangible output of the EIA process. Therefore, the effectiveness of the EIA system of a country depends largely on the quality of the EISs. The EIS is the product of the EIA process, and as such, the quality of the document is likely to be closely associated with the quality of the whole EIA process. This can be supported by the findings of the empirical study of Wende (2002) where the author shows that there is a clear relationship between the quality of EISs and the effectiveness of the EIA system.

A number of studies were conducted in both developed and developing countries on EIS quality (Glasson et al., 2005; Pinho et al., 2007; Sandham and Pretorius, 2008). These studies identified deficiencies in the contents of EISs, determined factors influencing the quality of EISs, and suggested improvements in the quality of EISs. Therefore, after the analysis of institutional requirements of an EIA system, it is imperative to look at the quality of EISs as a key aspect of the EIA system in Bangladesh.

2.6.3 Implementation of Mitigation Measures

An effective EIA system is also characterized by the proper implementation of mitigation measures along with monitoring and community participation. This is important as these post-EIS activities are an integral part of an effective EIA system. It is argued that adequate institutional arrangements and good quality of EISs alone cannot guarantee the protection of the environment. It is integral that the mitigation measures recommended by EIA reports are adequately implemented. The effective implementation of mitigation measures occur when all mitigation measures recommended by the

EIS are completely and adequately implemented. It is the effective implementation of the mitigation measures that occur at the post-EIS level that can make a project environmentally sustainable (the desired substantive outcome of an EIA).

There have been a number of studies undertaken in developed countries (for example, Ortolano & May) on the implementation of mitigation measures. There are a limited number of studies on the implementation of mitigation measures in developing countries, for example, Sanchez and Gallardo (2005) in Brazil, and Nadeem and Hameed (2010) in Pakistan. Therefore, the study of the post-EIS stage of the EIA process is a key area of this research.

2.6.4 Use of Criteria to Evaluate the EIA System

The application of criteria is becoming the prominent way of evaluation. Evaluation criteria are essentially a checklist of the requirements of an effective EIA system. These criteria are generally said to be normative provisions with which an EIA system should accord and they should be followed in practice. These good practice review criteria, as applied by a number of authors (such as Glasson and Salvador, 2000; Sadler, 1996), are well established and appear to be applicable to most jurisdictions. This study has used a set of criteria to evaluate the EIA system in Bangladesh.

2.7 EVALUATION CRITERIA

Many authors have been referencing good practice criteria to evaluate EIA systems, including Ahmad and Wood (2002), Glasson and Salvador (2000), and Sadler (1996). The criteria used by these authors reflect the requirements of an effective EIA system.

The application of these criteria is now widely recognized and established as the requirements of good EIA practice in evaluating EIA systems. The use of a set of criteria and the number of criteria depend on authors' own perspective and focus. Therefore, no set of criteria is exhaustive. With this in mind, a set of criteria (Appendix 2.1) was developed based on the above-mentioned authors to evaluate the EIA system in Bangladesh. Table 2.3 shows the summary of these criteria. In general, these criteria focus on the requirements and operation of the EIA system, including institutional arrangements, EIA procedure (quality of EISs), and the post-EIS stage of the EIA process. In this research, data were collected, analyzed, and interpreted in the light of these criteria.

Table 2.3 Summary of the criteria used to evaluate the EIA system in Bangladesh

Institutional arrangements

1. Clear legal basis of EIA system
2. Provisions of EIA requirements adequately prescribed by the legislation
3. Adequate technical and general EIA, SIA, and PP guidelines
4. Separate environmental agencies with adequate authority
5. Adequate resources and capacities of environmental agencies for implementing EIA in practice
6. Adequate interagency coordination between the environmental agency and other proponents
7. Other control mechanisms, such as an environmental court, code of conduct for EIA consultants

Quality of EISs

8. All relevant stages of EIA process are adequately addressed in practice
9. A competent authority is available to review and approve EISs
10. The EISs are able to be understood by all stakeholders
11. There are resources available to prepare good quality EISs (funds available, expert EIA consultants and adequate time for EIA study)
12. There is a code of conduct and accreditation system for EIA consultants to control the quality of EISs
13. The review of EISs is participatory and independent of an environmental agency

Implementation of mitigation measures

14. Recommended mitigation measures adequately implemented to address the predicted impacts of the projects
15. Adequate role is played by the environmental agency in successfully implementing the mitigation measures
16. Adequate monitoring activity during the implementation of mitigation measures
17. Active community participation during the implementation of mitigation measures
18. Other stakeholders (for example, donor agencies, Planning Commission) available to support the implementation of mitigation measures

Broader contextual factors

19. Political will favorable to environmental issues
20. Bureaucratic and developmental culture

Sources: Based on information from Sadler (1996) and Wood (1995).

2.8 EVOLUTION OF THE EIA SYSTEM IN BANGLADESH

Before independence in 1971, there was an environmental protection law titled *Water Pollution Control Ordinance* of 1970 (Government of East Pakistan, 1970). This ordinance on water pollution was the first legislation for the control, prevention, and abatement of the pollution of waters in East

Pakistan (renamed Bangladesh). A Pollution Control Board comprising the executive chiefs (secretaries) of concerned ministries was established to implement the ordinance. There was a provision for the punishment of noncompliant actors according to the ordinance. After the liberation of Bangladesh, this ordinance was rightly repealed and replaced by the *Environment Pollution Control Ordinance*, 1977 (GOB, 1977). This ordinance had broader scope than the previous one as it incorporated not only the control of water pollution but also other elements of environmental components such as air and soil. However, these pollution control laws did not mention the requirement of EIAs for development projects.

The evolution of the EIA dates back to the implementation of development projects in the 1960s in the water resource sector. Following the extensive floods in Bangladesh (East Pakistan) in 1954, a preventive strategy was taken by the government based on the recommendation of the *Krug Mission Report and East Pakistan Water and Power Development Authority* (EPWAPDA). The aim was to protect the agricultural lands against the average annual flood conditions in flood-prone areas through the construction of embankments along the major riverbanks and coastal development. This was the first massive intervention and construction of water resource development projects in Bangladesh.

After the liberation war and the independence of Bangladesh in 1971, the government endorsed this "modernization approach" and was quite uncritical of social and environmental aspects as there was no provision for EIAs. However, after the implementation of the projects, a number of studies were conducted by interested researchers to assess the environmental and social impacts of these projects in the early 1990s (Boyce, 1990; Brammer, 1990; Haque and Zaman, 1993). The findings of these studies revealed that these projects, as implemented in the 1970s and in the early 1980s, created economic benefits in the short term, but failed to ensure an environmentally and socially sound economy in the long run.

One of the projects, the Meghna-Dhonogoda project that was implemented in 1978 in the Chandpur District area, brought immense misery to thousands of people because of severe ecological, social, and cultural damage. The project caused a substantial reduction in fish production in the flood plains, thus adversely affecting the fishermen (Mirza and Erickson, 1996). There were other direct environmental and social impacts. Ultimately, the overall benefits of the projects were undermined by the extensive social and environmental costs. There was no provision of EIAs during the planning of these projects. This was a lesson for both the Government of Bangladesh and the donor agencies.

In the late 1980s, the first EIA was introduced on an *ad hoc* basis under the Flood Action Plan (FAP) (FAP-16 component) with the support of donor agencies, particularly the World Bank and the Asian Development Bank. Following the long flood in 1988, the Government of Bangladesh adopted multibillion dollar projects under the FAP to control and prevent floods. The development approach was, in fact, the continuation of a previous scheme that is known as Krug Mission's Master Plan for flood control. As there were adverse environmental and social impacts from the construction of embankments for flood control that occurred in the past, academics and environmentalists demanded that the negative environmental and social impacts of ensuing projects under the FAP had to be resolved at the planning stage.

In response to the recommendations made by academics and environmentalists from home and abroad, lessons learned from past experience, and pressure from donor agencies, the Government of Bangladesh incorporated EIAs to study the environmental and social impacts under the FAP. The first EIA guideline was prepared as part of the FAP-16 in 1992 to assess the ongoing and future FAPs and similar water management projects (FPCO, 1992). Since 1992, the EIA has been applied on an *ad hoc* basis (project by project) based particularly on the donor's demand. The application of EIAs was not mandatory for the development proponents until the enactment of the Environmental Conservation Act (ECA) in 1995.

In 1992, the Government of Bangladesh introduced the National Environmental Policy. In 1995, the government enacted the ECA making the application of EIA mandatory. The government prepared Environmental Conservation Rules (ECRs) in 1997 with detailed requirements for EIAs and thereby endorsed the ECA. In addition to donor support and domestic pressure from civil societies, such as the Bangladesh Centre for Advanced Studies (BCAS) and the International Union of Conservation of Nature (IUCN) office in Bangladesh, the Rio Declaration in 1992 influenced the government to enact formal EIA provisions.

2.9 STATUS OF THE EIA SYSTEM IN BANGLADESH

Current literature recognizes that there is no doubt about the role of EIA in making a project environmental friendly (Momtaz, 2005; Petts, 1999; Wood, 2003). Therefore, in order to harness the benefits of EIA, all developed countries and some developing countries (for example, South Africa, Brazil, Indonesia, Philippines, and China) have reviewed their EIA systems in detail and have developed better EIA systems. However, the role of EIA and the potential benefits of the EIA system in many developing countries including Bangladesh are still not well recognized or unknown. One of the reasons behind this is the lack of systematic scientific information about the EIA system.

Although the EIA was legally established in 1995 in Bangladesh, no systematic and comprehensive research has yet been done. A few attempts were made such as by Ahammed and Harvey (2004) and Momtaz (2002, 2003, 2005) to understand the status of the EIA in Bangladesh. These studies focus predominantly on institutional (and to some extent procedural) issues and, therefore, provide only a partial view of the EIA system. According to Momtaz (2002), in order to understand the effectiveness of an EIA at full scale, a thorough understanding of the relationship between the EIA and the environmental protection assurance in the context of Bangladesh is necessary (Momtaz, 2002, p. 178).

Furthermore, there is no provision of routine assessment of EIA effectiveness in Bangladesh at the government level. In many countries (for example, the EIA Directive in EU countries), the EIA is regularly reviewed and the legislation, the administrative setup, and the ways of practice are modified. In Bangladesh, due to the lack of detailed scientific information, no significant improvement of the EIA has been made since 1995. Notably, in the absence of a detailed study, we still do not fully understand whether the EIA system in Bangladesh is complying with good practice and fulfilling its intended objectives. Therefore, it is important to examine the EIA systems more closely and comprehensively. This will help to identify the actual barriers and realistic opportunities for the optimization of this management tool.

2.10 CONCLUSION

The aim of this chapter was to review the existing literature on EIA effectiveness and to develop a conceptual framework (integrated holistic approach) to evaluate the EIA system in developing countries with special focus on Bangladesh. In this chapter, existing approaches to the evaluation of an EIA system were analyzed and a conceptual holistic framework was developed to evaluate the EIA system in Bangladesh. Finally, the evolution and status of the EIA in Bangladesh was discussed.

APPENDIX

Appendix 2.1 Checklist/Criteria
Checklist for Review of EIA System in Bangladesh

1. Clear legal arrangements for effective EIA system
   a. Clear legal basis of EIA system
   b. Clearly documented act and regulations for EIA process
2. Adequate legal arrangements for EIA system
   a. Necessary steps of EIA process (from screening to monitoring and implementation of mitigation measures) adequately mentioned by act and regulations as mandatory
   b. Procedural requirements of EIA are clearly mentioned by legislations and enforceable in practice
   c. Clear outline of EIA procedures with roles and responsibilities of stakeholders and time limits exist for EIA system
   d. EIA legislation covers both government and private projects applicable to EIA
   e. EIA requirements are enforceable through the court's action
   f. EIA legislation clearly specifies community participation at all stages of EIA process
   g. The legislated EIA includes broad definition of environmental/ coverage of factors
   h. Overall, the legal provisions of an effective EIA system are adequate and clearly understandable and enforceable in practice
3. Are there a set of adequate local EIA guidelines and manuals in place for the following:
   a. Guidelines sector-specific projects
   b. EIA report preparation and contents
   c. For EIA report review
   d. Prediction and evaluation of impacts
   e. Public participation in EIA process
   f. Development of mitigation action plan and its implementation
4. International environmental policies, agreements, and conventions exist in place to support EIA
5. The EIA is integrated into project cycle and integration of environmental factors into project design, approval, and implementation occurs properly
6. There are codes of conduct for controlling EIA consultants
7. Accreditation system to recognize genuine and skilled EIA consultants

Appendix 2.1 Checklist/Criteria—cont'd

8. Adequate administrative arrangements for EIA system
  a. Separate environmental agency responsible for administration and development of EIA system
  b. Sufficient enforcement authority and capacity of environmental agency (DPE) to administer and implement EIA
  c. The environmental agency is appropriately decentralized for EIA activities
  d. Adequate staff for EIA unit in the environmental agency (DOE) to implement EIA
  e. Staff with adequate skills at both central and lower tiers for EIA administration
  f. Adequate measures in place for human resource development with adequate training programs
  g. Stable leadership and incentives for staffs of environmental agency
9. Regular sufficient financial and logistic support for the environmental agency from government
10. Effective interagency coordination mechanism
  a. Coordination between the environmental agency and other stakeholders including sectoral agencies
  b. EIA or environmental cell in all relevant agencies related to development and resource management whose activities bear environmental cost
  c. The position of environmental agency in the government hierarchy is strong
  d. Adequate support from other stakeholders (donor agencies, Planning Commission) to implement EIA during the hole life cycle of the project
11. Methodological requirements (preparation of EISs), review, and approval of EISs
  a. Screening process/triggers of EIA
    a1. There is a legal test of whether the action likely affects the environment significantly
    a2. There is a clear and updated checklist of the type of action to be subject to EIA
    a3. Clear criteria/thresholds exist for all sectoral projects
  b. Scoping of EIA
    b1. Proponents consult stakeholders and the environmental authority early in the EIA process
    b2. Action-specific scoping guidelines and TOR are prepared by the proponent

(Continued)

Appendix 2.1 Checklist/Criteria—cont'd

b3. Overall, the scoping involves identification of key environmental components, mode of EIA study, baseline data and clear role, and responsibilities that EIA studies to facilitate detail EIA

c. Impact identification and assessment of impacts

c1. The contents of EIA reports cover all significant impacts

c2. Best available methods or techniques employed for impact assessment

c3. Accreditation system and code of ethics for EIA consultants exist and effectively applied

c4. Environmental agency facilitates proponents to prepare EISs providing with necessary information and consultation

d. Alternative options are sufficiently evaluated to minimize environmental impacts of projects

e. Environmental management plan sufficiently includes mitigation measures and monitoring activity

12. Review of EIA reports and environmental clearance

a. An effective review system for EIS review is in place

b. An independent review body with appropriate expertise exist

c. Checks (checklist or review criteria, TOR) on the objectivity of the EIA report review exist and employed

d. Review of EIA reports well recorded and published by the environmental agency

13. Community involvement in EIA process

a. Community involvement in EIA process is specified by EIA law

b. Community have opportunity to participate in the whole EIA process

c. Methods/tools used for communitarian and involvement appropriate for all types of stakeholders

d. Public inputs formally taken into account by the proponents during the design and implementation of the project

14. Implementation of mitigation measures (post-EIS stage of EIA)

a. Proponents implement the project in accordance with the conditions of approval of EIA reports

a1. Mitigation measures recommended by the EIA report implemented by the proponents adequately

a2. Proponents involve community actively in the implementation of mitigation measures

a3. Role of important stakeholders such as donor agencies is in place during the implementation of mitigation measures and evaluate the environmental performance of the project

a4. Adequate fund for the implementation of mitigation measures

Appendix 2.1 Checklist/Criteria—cont'd

a5. Adequate support and supervision are in place from the environmental agency during the implementation of mitigation measures

a6. There is an internal coordination (between EIA team and project management team) within the proponent's management

15. Formal EIA monitoring in place during the implementation of mitigation measures and operation of projects

a. Supervision and monitoring activities are carried out by the environmental agency to make the implementation of mitigation measures effective

b. Adequate supervision and monitoring program are in place and carried out by the proponents

c. Proponent regularly submits monitoring results to the environmental agency

d. Proponents share monitoring results and information on mitigation with community

e. Community people have active participation in monitoring activity

16. Lessons learned from implementation of mitigation measures are incorporated by policymakers and proponents for further development of EIA

17. Environmental agency impose penalties/sanctions against public and private proponents for noncompliance with approval decisions

18. Adequate broader contextual support exits for effective implementation of EIA process

a. Environmental governance

b. Adequate international environmental laws and conventions are in place and their implementation

c. Bureaucratic culture and the mode of economic development

d. Political will

e. Socioeconomic conditions

REFERENCES

Abaza H, Bisset R, Sadler B. Environmental impact assessment and strategic environmental assessment: towards an integral approach. Geneva, Switzerland: United Nations Environmental Program; 2004.

Abracosa R, Ortolano L. Environmental impact assessment in the Philippines: 1977–1985. Environ Impact Assess Rev 1987;7(4):293–310.

Ahammed AKMR, Harvey N. Evaluation of environmental impact assessment procedures and practice in Bangladesh. Impact Assess Project Appraisal 2004;22(1):63–78.

Ahmad B, Wood C. A comparative evaluation of the EIA systems in Egypt, Turkey and Tunisia. Environ Impact Assess Rev 2002;22:213–34.

Appiah-Opoku S. Environmental Impact Assessment in developing countries: the case of Ghana. Environ Impact Assess Rev 2001;21:59–71.

Bina R. Context and systems: thinking more broadly about effectiveness in strategic environmental assessment in China. Environ Manage 2007;42:717–33.

Blackmore R, Wood C, Jones CE. The effects of environmental assessment on UK infrastructure project planning decisions. Plan Prac Res 1997;12(3):223–38.

Boyce JK. Birth of a megaproject: political economy of flood control in Bangladesh. Environ Manage 1990;14(4):419–28.

Brammer H. Floods in Bangladesh: flood mitigation and environmental aspects. Geogr J 1990;156(2):158–65.

Briffett C. Environmental impact assessment in East Asia. In: Petts J, editor. Handbook of environmental impact assessment: environmental impact assessment in practice: impact and limitations, vol. 2. Oxford: Blackwell; 1999. p. 143–67.

Brown AL, Hindmarsh RA, McDonald GT. Environmental assessment process and issues in the Pacific Basin-Southeast Asia Region. Environ Impact Assess Rev 1991;11:143–56.

Cashmore M, Gwilliam R, Morgan R, Cobb D, Bond A. The interminable issue of effectiveness: substantive purposes, outcomes and research challenges in the advancement of environmental impact assessment theory. Impact Assess Project Appraisal 2004; 22(4):295–310.

Cashmore M, Bond A, Cobb D. The contribution of environmental assessment to sustainable development: toward a richer empirical understanding. Environ Manage 2007;40: 516–30.

Doberstein B. Environmental capacity-building in a transitional economy: the emergence of the EIA capacity in Vietnam. Impact Assess Project Appraisal 2003;21(1):25–42.

Doyle D, Sadler B. Environmental impact assessment in Canada: frameworks, procedures & attributes of effectiveness. A report in support of the international study of the effectiveness of environmental assessment, Canada; 1996.

Ebisemiju FS. Environmental impact assessment: making it work in developing countries. J Environ Manage 1993;38:247–73.

FPCO. Fisheries studies and pilot projects. Flood Plan Coordination Organization Flood Action Plan (FAP)-17. Final report (draft) with supporting volumes. Dhaka; 1992.

Glasson J, Salvador NNB. EIA in Brazil: a procedure-practice gap. A comparative study with reference to the European Union, and especially the UK. Environ Impact Assess Rev 2000;20:191–225.

Glasson J, Therivel R, Chadwick A. Introduction to environmental impact assessment. London: Taylor and Francis Group; 2005.

Government of Bangladesh. The Environmental Pollution Control Ordinance (Ordinance No XIII of 1977). Dhaka; 1977. p. 1–6.

Government of East Pakistan. Pollution Control Ordinance. East Pakistan Ordinance, No. V of 1970; 1970.

Haque CE, Zaman MQ. Human response to Riverina hazards in Bangladesh: a proposal for sustainable floodplain development. World Dev 1993;21(1):93–107.

Kakonge JO, Imevbore M. Constraints on implementing environmental impact assessments in Africa. Environ Impact Assess Rev 1993;13:299–308.

Lawrence DP. The need for EIA theory-building. Environ Impact Assess Rev 1997;17:79–107.

Lee N, George C, editors. Environmental assessment in developing and transitional countries: principles, methods and practice. New York: John Wiley & Sons, Ltd.; 2000.

Lim G. Theory and practice of EIA implementation: a comparative study of three developing countries. Environ Impact Assess Rev 1985;5:133–53.

Lohani BN, Evans JW, Everitt RR, Ludwig H, Carpenter RA, Tu S. Environmental impact assessment for developing countries in Asia. vol. 1. Manila: Asian Development Bank; 1997 p. 1–356.

Mirza MQ, Erickson NJ. Impact of water control projects on fisheries resources in Bangladesh. Environ Manage 1996;20(4):523–39.

Momtaz S. Environmental impact assessment in Bangladesh: a critical review. Environ Impact Assess Rev 2002;22:163–79.

Momtaz S. The practice of social impact assessment in a developing country: the case of environmental and social impact assessment of Khulna-Jessore Drainage Rehabilitation Project in Bangladesh. Impact Assess Project Appraisal 2003;21(2):125–32.

Momtaz S. Institutionalizing social impact assessment in Bangladesh resource management: limitations and opportunities. Environ Impact Assess Rev 2005;25:33–45.

Nadeem O, Hameed R. Evaluation of environmental impact assessment in Pakistan. Environ Impact Assess Rev 2008;28(8):562–71.

Nadeem O, Hameed R. Exploring the potential and constraints to implementing the international best practice principles of EIA follow-up: the case of Pakistan. J Am Sci 2010; 6(12):108–21.

Nooteboom S. Impact assessment procedures for sustainable development: a complexity theory perspective. Environ Impact Assess Rev 2007;27:645–65.

Odum EP. The emergence of ecology as a new integrative discipline. Science 1977; 195(4284):189–1295.

Ortolano L, Jenkins B, Abracosa RP. Speculations on when and why EIA is effective. Environ Impact Assess Rev 1987;7:285–92.

Owens S, Rayner T, Bina O. New agendas for appraisal: reflections on theory, practice, and research. Environ Plan 2004;36:1943–59.

Paliwal R. Environmental impact assessment practice in India and its evaluation using SWOT analysis. Environ Impact Assess Rev 2006;26(5):492–510.

Petts J. Introduction to environmental impact assessment in practice: fulfilled potential or wasted opportunity? In: Petts J, editor. Handbook of environmental impact assessment: environmental impact assessment in practice: impact and limitations, vol. 2. Oxford: Blackwell; 1999.

Pinho P, Maia R, Monterroso A. The quality of Portuguese environmental impact studies: the case of small hydropower projects. Environ Impact Assess Rev 2007;27:189–205.

Ross WA. Environmental impact assessment in the Philippines: progress, problems and direction for the future. Environ Impact Assess Rev 1994;14(4):217–32.

Sadler B. International study of the effectiveness of environmental assessment: final report. Canadian Environmental Assessment Agency/International Association for Impact Assessment; 1996.

Sanchez LE, Gallardo ALCF. On the successful implementation of mitigation measures. Impact Assess Project Appraisal 2005;23(3):182–90.

Sandham LA, Pretorius HM. A review of EIA report quality in the North West province of South Africa. Environ Impact Assess Rev 2008;28(4–5):229–40.

Tongcumpou C, Harvey N. Implications of recent environmental impact assessment changes in Thailand. Environ Impact Assess Rev 1994;14(4):271–94.

Turnbull J. Environmental impact assessment in the Fijian state sector. Environ Impact Assess Rev 2003;23(1):73–89.

Vidyaratne H. Environmental impact assessment theories and practice: balancing conservation and development in Sri Lanka. J Environ Assess Policy Manage 2006;8(2):205–22.

Wang Y, Morgan RK, Cashmore M. Environmental impact assessment of projects in the People's Republic of China: new law, old problem. Environ Impact Assess Rev 2003;23(5):543–79.

Wende W. Evaluation of the effectiveness and quality of environmental impact assessment in the Federal Republic of Germany. Impact Assess Proj Appraisal 2002;20(2):93–99.

Wood C. Environmental impact assessment—a comparative review. London: Prentice Hall; 1995.

Wood C. Environmental impact assessment in developing countries. Int Dev Plan Rev 2003;25(3):301–21.

Zeremariam TK, Quinn N. An evaluation of environmental impact assessment in Eritrea. Impact Assess Project Appraisal 2007;25(1):53–63.

Zubair L. Challenges for environmental impact assessment in Sri Lanka. Environ Impact Assess Rev 2001;21:469–78.

CHAPTER *3*

Institutional Framework of the Environmental Impact Assessment System

Contents

3.1 INTRODUCTION

Institutional arrangements (legal and administrative) for environmental impact assessment (EIA) act as the foundation for EIA practice in a country

Evaluating Environmental and Social Impact Assessment in Developing Countries   © 2013 Elsevier Inc.
http://dx.doi.org/10.1016/B978-0-12-408129-1.00003-6   All rights reserved.

and show the country's commitment in protecting the environment from the potential impacts of development projects. The environmental policies, laws, regulations that define a country's legal framework for EIA system need to be well organized, clear, and comprehensive so that the EIA can work effectively to achieve its goals. In addition to a strong and clear legal support, adequate administrative arrangement is also a prerequisite for an effective EIA system.

A substantial body of literature on institutional arrangement of EIA system is available both in developed and in developing countries (Abaza et al., 2004; Goodland and Edmondson, 1994; Modak and Biswas, 1999). These studies show that both developed and developing countries have been continuously reviewing their legal and administrative arrangements of EIA. Legal amendments and administrative reforms are made to strengthen the foundation of EIA systems and increase their scope and effectiveness. Developed countries introduced EIA earlier than the developing countries and have reviewed and established adequate legal and administrative support for effective EIA practice. For example, there has been a comprehensive reform of long-established EIA laws in New Zealand (1991), Canada (1995), and Australia (1997). The EU directive is continuously modified and reviewed every 5 years. Although developing countries introduced EIA at a later date, many developing countries such as China (2001), Thailand (1992), The Philippines (1992), and India (1991) have already reviewed their EIA systems and replaced the old ones with improved versions with subsequent reorganization of administrative arrangements.

In Bangladesh, EIA has been legally functional since the enactment of Environmental Conservation Act (ECA) in 1995 (DOE, 1995). There have been some studies (for example, Ahammed and Harvey, 2004) on EIA system in Bangladesh. However, there have been no in-depth studies to particularly understand whether the institutional arrangements are adequate for effective EIA practice.

3.2 LEGAL ARRANGEMENTS FOR THE EIA SYSTEM

Legal arrangements for the EIA system in Bangladesh broadly include constitutional rights, national environmental policy, plan, strategy, and, more importantly, the ECA and Environmental Conservation Rules (ECR) (Figure 3.1). These are described below.

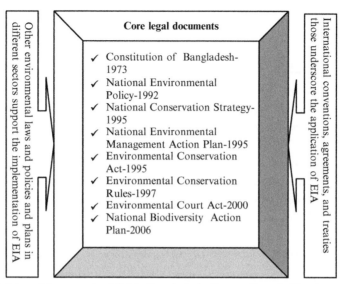

Figure 3.1 Legal framework of EIA in Bangladesh. *Source: Compiled by the authors.*

3.2.1 The Constitution of Bangladesh

The constitution of Bangladesh protects "the right to life and personal liberty" (Articles 31 and 32) as a fundamental human right (Government of Bangladesh, 1973, p. 43). Although it does not explicitly recognize the rights to clean environment as a fundamental right, in two cases (XVII Bangladesh Legal Digest, 1996 (AD) p. 1 and XLVIII DLR, 1996, p. 438) related to the protection of environmental impact of development projects, the Supreme Court of Bangladesh has resolved that the "right to life" enshrines as a fundamental right including the "right to a healthy environment." Thus, the right to environmental protection based on these historic judgements is now constitutionally supported in Bangladesh.

3.2.2 National Environmental Policy

In response to the United Nations Conference on Environment and Development 1992, the Government of Bangladesh (GOB) approved a cross-sectoral policy, the National Environmental Policy (NEP) in 1992 (Ministry of Environment and Forest (MoEF), 1992). The NEP reiterates a broad-based approach to environment and development that was embodied in Rio Declaration. The NEP is predicated on a growing concern for the degradation of natural resource base in Bangladesh and the potential consequences of this degradation. It also takes into account how the natural resource base is affected by the development activities of different sectors.

3.2.3 National Environmental Management Action Plan

The National Environmental Management Action Plan (NEMAP) of 1995 has been formulated with the objective to identify environmental concerns in various sectors and consequent actions to address the environmental concerns. NEMAP is considered as the basis for concretizing programs and interventions aimed at promoting better management of scarce resources and reversing the trend of environmental degradation. The NEMAP goes through all concerned sectors and identifies the areas of actions to promote sustainable development in Bangladesh.

3.2.4 The National Conservation Strategy

The National Conservation Strategy (NCS) was prepared in 1995 in response to the Global Conservation Strategy of the International Union of Conservation of Nature in early 1990s. It is an important step toward achieving the objectives and integrating the policies on environment. By adopting the NCS, the government not only reinforces its national and international commitments toward sustainable development but also gives commitment for effective application of EIA.

3.2.5 National Biodiversity Action Plan

Furthermore, in 2006, the GOB prepared National Biodiversity Action Plan in response to the International Convention of Biological Diversity. The major objective of this document is to conserve and restore biodiversity of the country for the well-being of the current and future generations. One of the key strategies of conserving the biodiversity is the use of EIA to integrate relevant biological issues to protect the country's biodiversity.

3.2.6 Environmental Conservation Act

The first major law for the protection and conservation of nature is ECA of 1995. The ECA '95 provides the legal basis for EIA in Bangladesh. The ECA calls for the assessment of environmental impact of new development projects. Section 12 of ECA stipulates that "no industrial unit or project shall be established or undertaken without obtaining an Environmental Clearance from the DOE in the manner prescribed by the rules." Section 15(8) of ECA declares punitive action against the violation of Section 12.

The Act also stipulates the actions for the enforcement of EIA at post–EIS stage. There are penalties specified by ECA when a project damages the eco-system or pollutes the environment during its implementation or operation. The DOE can take administrative action including determination of com-pensation. It can file compensation case or criminal case at an environment court (Section 7) against the polluters.

3.2.7 Environmental Conservation Rules

The ECR of 1997 have been promulgated for the implementation of ECA where the ECR establish a greater detail on the requirements of EIA pro-cedure and implementation (Table 3.1). The ECR explicitly prescribe the EIA requirements with particular focus on the process of Environmental Clearance. According to ECR '97, all types of environmentally significant projects will be subject to EIA and the EIA reports will need to be submitted to DOE for approval. The ECR also stipulate the responsibilities of DOE to implement, coordinate, and enforce the EIA provisions properly during an EIA study and approve the EIA report. The sections from 7 to 14 of ECR

Table 3.1 Major features of Environmental Conservation Act (1995) and the Environmental Conservation Rules (1997) as legal foundation of EIA

Environmental Conservation Act (1995)
1. Assistance from law enforcing agencies
2. Polluters pay principle
3. Formulation of environmental guidelines
4. Claim for compensation by the affected people
5. Penalties for violation of project approval conditions
6. Offenses committed by proponents and action against proponents
7. Power to make rules

Environmental Conservation Rules (1997)
1. Complain by community to DG against non-compliance of project approval conditions
2. Power to collection of sample
3. ECC procedure with specified time limit
4. Screening of both public and private projects
5. Review of EIS
6. No objection Certificate from Local Government
7. Appeal by proponents against order or directions from DOE
8. Determination of environmental standards

Sources: Based on information from Environmental Conservation Act (1995) and Environmental Conservation Rules (1997).

describe the process of environmental clearance, the roles and responsibilities of DOE, validity of Environmental Clearance Certificate (ECC), and procedure of appeal and hearing about the rejection of EIA report by the DOE (DOE 1997a, p. 183–8).

3.2.8 International Conventions, Treaties, and Protocols

International environmental laws and policies have implications for the EIA systems of the countries that sign or endorse them. These nonbinding (such as Rio Declaration on Sustainable Development) and legally binding agreements and conventions establish important principles and carry various obligations for signatory countries. The agreements signed and ratified by the GOB are shown in Table 3.2.

Although legislations often play important roles in environmental governance, they do not necessarily guarantee that the intents of the legislators will be materialized unless there is a favorable administrative arrangement to enforce them.

3.3 ADMINISTRATIVE/ORGANIZATIONAL FRAMEWORK OF EIA

The administrative framework for EIA is determined by the variety of participants, their respective roles, and their interactions. Generally, the EIA system involves public and donor agencies, consultants, review agency,

Table 3.2 International conventions, treaties, and protocols relevant to EIA in Bangladesh

Agreements related to the conservation of nature and natural resources include the following:
- Convention on International Trade in Endangered Species of Wild Fauna and Flora (CITES) signed in 1973
- Convention on Biological Diversity (CBD) signed in 1992 and ratified in 1994

Agreements related to the prevention and controls of pollution include the following:
- Framework Convention on Climate Change (FCCC) signed in 1992 and ratified in 1994
- Convention for the Protection of Ozone Layer and Montreal Protocol on Ozone Depleting Substances signed 1992 and ratified in 1994
- Basel Convention on the Control of Trans-boundary Movements of Hazardous Wastes and their Disposal signed and ratified in 1993

Source: Compiled by the authors.

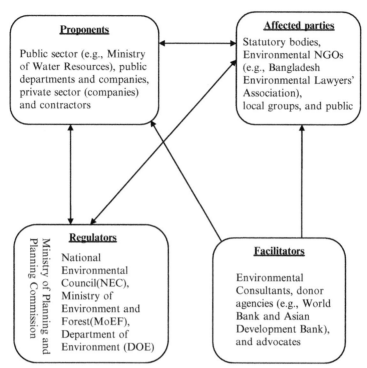

Figure 3.2 Key actors involved in EIA system in Bangladesh and their interaction. *Source: Compiled by the authors.*

NGOs, media, and judicial agencies. Figure 3.2 shows the interactions between these actors. The core government agencies directly involved in administering EIA are described below.

3.3.1 Ministry of Environment and Forest

There was no separate or single authority to deal exclusively with environmental issues in Bangladesh until 1989. This instructional constraint impeded the optimum use of resources as each sectoral agency pursued its own objectives with little coordination and integration. The GOB created MoEF in 1989 for planning, promoting, coordinating, and overseeing the implementation of environmental programs.

The MoEF bears the responsibility for coordinating and working with other ministries at national level to ensure that environmental concerns are given due recognition in their development programs and projects. The MoEF has direct authority to supervise and guide the activities of its subordinate departments including the Department of Environment

(DOE) and formulate policy and rules when and where necessary. The MoEF is a permanent member of the Executive Committee of the National Economic Council (ECNEC), the highest decision making body for economic policy issues and the highest authority to approve all public investment projects.

3.3.2 National Environmental Committee

The GOB established a broad-based National Environmental Committee (NEC) to provide overall guidance in the implementation of NEP. It is the highest coordinating and advisory body for environmental management and development with the direct support from the Cabinet Division, Office of the Prime Minister, and the MoEF. As a multiagency body, the NEC is intended to serve a unifying role involving facilitating coordination among the sectoral agencies and creating awareness among stakeholders about effective implementation of EIA.

3.3.3 Planning Commission

Planning Commission is the central planning agency within the Ministry of Planning. It guides the country's development strategy. It is in charge of the preparation and funding of Five-Year Plans under which development programs and projects are formulated across sectors.

3.3.4 Department of Environment

The DOE is the responsible authority to implement and enforce EIA within the MoEF. It has the responsibility to formulate EIA guidelines for different sectors and review and approve EIA reports. It has the authority to prosecute against any agency or persons who are polluting the environment and breaching the environmental directions served by the DOE, thereby DOE has been given extensive power to deal with environmental pollution across the sectors. In addition to EIA-related functions, the mandate of DOE is broadly to ensure conservation of environment, assessment, and improvement of environmental standards and monitoring and mitigation of pollution control.

The DOE is headed by a Director General (DG) equivalent to the position of Additional Secretary in civil bureaucracy in Bangladesh. The responsibilities of DG are discharged through the Head Office located in Dhaka and six divisional offices all over Bangladesh.

There are no offices of DOE at district or Upa–Zilla (subdistrict) levels, although there are almost all other government offices functioning at both districts and Upa–Zilla levels. Within the office of the Director (technical) at the head office, there is an office (unit) headed by a Deputy Director (EIA) to coordinate and discharge EIA-related services.

3.4 EIA PROCESS AND ADMINISTRATIVE PROCEDURE

According to the ECA and ECR, the administrative procedure of EIAs is controlled and managed by the DOE. A tiered approach is adopted in the assessment of environmental impacts of projects. EIA procedure passes through the following three tiers—screening, initial environmental examination (IEE), and EIA.

3.4.1 Screening

Screening decides whether the EIA process should be applied to a development, and if it is required, its type, that is, IEE or EIA. The screening is based on several criteria such as type of project, its size, and location. Industrial projects have been divided into four categories for the purpose of screening— Green, Amber A, Amber B, and Red (DOE, 1997a), according to their environmental significance and location. Category Green projects do not require IEE and EIA. A no objection certificate (NOC) from the local authority is adequate for a project that falls into this category. On the other end of the range are the Red category projects that require both IEE and EIA. This normative screening process enables the DOE and the proponents to determine which steps to follow in the environmental clearance process (Figure 3.3).

The DOE has developed a list of all major industries/activities under all these categories. This list helps proponents to go through the screening process to determine which category their proposal falls under. This normative screening process facilitates the process of clearance from the DOE. Based on the category of their proposed activity, the proponents then go through certain steps to get approval from the DOE (Figure 3.3).

3.4.2 Initial Environmental Examination

A typical IEE needs to provide the following information: description of the project, description of the existing background environment around the

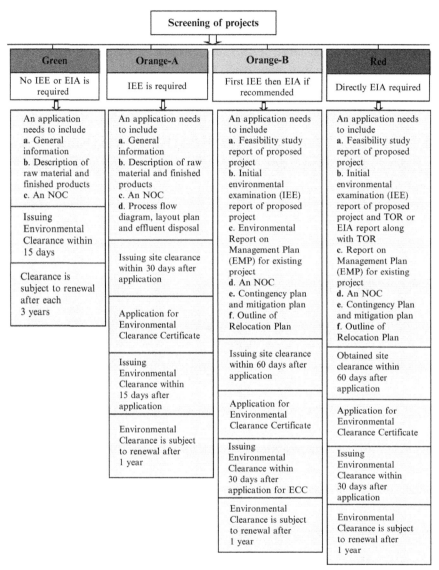

Figure 3.3 Screening and environmental clearance process by the DOE. (For color version of this figure, the reader is referred to the online version of this chapter.) *Source: DOE (1997b).*

project site (generally this should cover an area of 1-km radius), potential significant impacts (both during construction and operational phases), mitigative and abatement measures, residual impacts if any, and monitoring program.

3.4.3 Detailed EIA

For the projects where environmental issues are not fully addressed in IEE because of lack of information or due to the higher magnitude impacts, an EIA is required. The EIA study is to focus on addressing issues which remained unresolved in the IEE. The steps involved in conducting an EIA are similar to those followed elsewhere around the developed and developing world. They are described below:

a. *baseline studies* are similar to scoping that is conducted to identify issues and affected interest groups and to determine study boundaries. They usually involve site selection and effects, material to be used, storage and transport, machinery and equipment, and requirement of utilities;

b. *impact identification* involves compilation of a list of key sources of impacts of the project on environment and study in detail the sources of impacts of the project such as emissions, water consumption, wastewater generated, noise generated, etc;

c. *evaluation of the significance of the impacts*, i.e., whether the impacts are positive or negative, long term or short term, significant or insignificant, reversible or irreversible, avoidable or unavoidable, and the determination whether mitigation measures are required;

d. *development of mitigation measures* through change in project site, processes or raw materials; changes to operating methods, disposal routes or locations; and changes to engineering design and methods of construction; and

e. a *monitoring plan* would need to be developed and presented as a part of a Management Plan. The Management Plan will also outline how mitigation measures are going to be implemented with clear indications of responsible officers.

3.5 THE METHODOLOGIES OF IMPACT IDENTIFICATION AND ASSESSMENT

Various methods and techniques have been used to identify impacts of proposals/actions/projects on the biophysical and socioeconomic environment and to make assessment and evaluation of their magnitude and significance. These are *ad hoc*, checklists, matrices, networks, overlays, environmental index using factor analysis, cost–benefit analysis, simulation modeling workshops, expert opinion, etc. Of these, only a few have survived the test of time. Use of the others has become restrictive or obsolete due to their technical nature and very limited use. The ones that survived the test of time due to their usefulness, simplicity, and ease of use are checklists, networks, and matrices. Overlay methodology that was proposed by McHarg in the 1970s (McHarg, 1969) as a manual process of inputting spatial data on maps and overlay various map transparencies to visualize extent of impacts on vegetation, soil, water (social values), etc. has evolved into a very sophisticated visual technique supported by geographic information system. In Bangladesh, DOE has proposed the use of checklist and matrix methodologies. It has developed "checklist of environmentally and otherwise sensitive areas," that is, "checklist of typical activities for some projects" including investigation stage, site preparation and construction, operation and maintenance, and future and related activities; and "checklist of environmental components" including physicochemical, biological, and human. It has also provided a long explanation of environmental characteristics and components for the proponents (DOE, 1997b; EIA Guidelines for Industries, p. 51, 52, 54, 58).

3.6 REVIEW OF EIA REPORTS AND ISSUANCE OF ECC

EIA reports are reviewed by the review committee in DOE. The committee consists of the DOE officials and is headed by the DG of DOE. However, DOE contracts EIA experts and academics from outside as members of the review committee for any special cases on *ad hoc* basis. Based on the review reports, the DG of DOE accepts the EIS, asks for more information, or rejects the EIS (Figure 3.4).

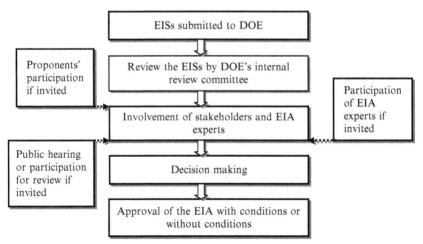

Figure 3.4 EIS review process in Bangladesh. *Source: Prepared by the authors based on discussions with DOE officials and information from DOE (1997b).*

If accepted, the DOE issues an ECC to the proponents outlining terms and conditions. The terms and conditions usually involve that the proponents have to take adequate and appropriate mitigation measures as designed in the Environmental Management Plan (EMP) and implement the measures properly during the implementation and operation of the project. It is the responsibility of the proponents to arrange and monitor the implementation of mitigation activities and report to the DOE regularly. The DOE is responsible to check regularly if the proponent is implementing the EMP properly. After the implementation of the project, a final evaluation (environmental auditing) of the EIA is undertaken jointly by the DOE and the proponent.

3.7 PENALTIES FOR NONCOMPLIANCE OF ECC

The DOE is empowered to take punitive action against the noncompliant proponents. Initially, the DOE sends notice to the proponent as a warning. If the proponent does not rectify the problem after the notice, the DOE directs the utility service departments such as electricity, gas, and telephone service authorities to stop all services. According to the ECR '97, the DOE is authorized to fine the noncompliant proponent. The organization also has the authority to file cases in the courts against the proponents. If the offenses are proved, the proponents may be jailed for a maximum of 7 years and fined as well.

There are many other governmental and nongovernmental agencies and international development partners directly or indirectly involved in the conduct of EIA. The most important among them are Local Government Engineering Department (LGED), Centre for Environmental and Geographic Information Systems, Water Resources Planning Organization, CARE Bangladesh, Asian Development Bank (ADB), and World Bank (WB). These organizations have their own EIA processes and conduct EIA of the projects supported and/or funded by them. The processes are similar to the ones proposed by the DOE.

3.8 STRENGTHS AND WEAKNESSES OF INSTITUTIONAL ARRANGEMENTS OF EIA IN BANGLADESH

This study identifies both strengths and weaknesses of institutional arrangements of EIA (both legal and organizational) as shown in Table 3.3.

Table 3.3 Key strengths and weaknesses of institutional arrangements of EIA in Bangladesh

Strengths

  Legislative

  - ✓ Clear legal basis for EIA system (ECA and ECR)
  - ✓ Constitutional support for environmental protection
  - ✓ Broader definition of environment
  - ✓ Clear list of projects for screening
  - ✓ Clear timeline for approval of EIS and issuance of ECC
  - ✓ Provisions for penalties

Organizational

  - ✓ Separate environmental agency
  - ✓ Clear and adequate power of DOE to implement EIA
  - ✓ Incorporation of EIA requirements in national economic development plans and sectoral policies
  - ✓ Simultaneous occurrence of EIA with feasibility study

Weaknesses

  Legislative

  - ✓ The ECC needs to be renewed every year
  - ✓ Provision of site clearance undermine the effective review of EIA report
  - ✓ No provision for expansion of new project and renovation of old project
  - ✓ No explicit provision is made for undertaking all stages of EIA
  - ✓ No provision for affected community to directly go to court
  - ✓ Environmental quality standards are not up-to-date

Organizational

  - ✓ Inadequate manpower and budget
  - ✓ Leadership crisis and lack of incentives for DOE staff
  - ✓ Inadequate interagency cooperation at national and local level

---

Source: Compiled by the authors from NVivo analysis.

3.8.1 Strengths of the Legal Provisions

a. *Clear legal basis of EIA*: The study shows that a well-defined and specific legal provision of EIA exists in Bangladesh. The ECA of 1995 and the ECR of 1997 are the milestones for the country in terms of preserving the environment through the application of EIA. The ECA and the ECR have clearly specified the responsibilities of the DOE and the proponents to carry out the EIA.

b. *Clear list of projects and screening*: According to the ECR of 1997, there is a classification of development projects based on specific criteria (such as size, type, and location) for projects to be screened. Because of the presence of a clear list of projects, proponents do not require decisions from the DOE as to whether the intended project requires IEE or full EIA.

c. *Broader scope of EIA application*: According to The ECA '95, "environment means the interrelationships existing between water, air, soil, and physical property and their relationship with human beings, other animals, plants, and microorganisms" (DOE, 1995, p. 155). Thus, the definition recognizes not only environmental issues but also social issues.

d. *Involvement of local government into EIA process*: Although EIA is undertaken by proponents and reviewed centrally by the DOE, there is a provision that the proponent has to obtain an NOC from the relevant local government (DOE, 1997a). This shows a strong commitment of the government to democratize the EIA process through the involvement of elected local government representatives. Thus, the provision of NOC enforces proponents to involve local people from the beginning of the project planning.

e. *Public environmental litigation and effective judicial control*: Before 1996, the environmental laws allowed only the grieved person or the community affected by a development project to go to court to get actions against polluters. After the historic judgment of Bangladesh Supreme Court in 1996, any person or organization can go to court to sue the polluter on behalf of the community. With this achievement, some organizations such as Bangladesh Environmental Lawyers Association (BELA) have been encouraged to act against any development activities subversive to environmental quality. For example, the recent lawsuits and successes by BELA against the government for not implementing EIA, the Tannery Industries in Dhaka city for releasing toxins, or the Ship Breaking Industries in Chittagong for not adopting and implementing EMP show that the court is sensitive and sympathetic about environmental issues.

f. *Environmental quality parameters*: The effectiveness of EIA process in protecting the environment also depends on the degree of environmental protection offered by the standards. In fact, the formulation of environmental quality parameters serves as the basis for planning of development projects particularly for designing Environment Management Plan. The environmental quality standards set by the DOE include both ambient environmental standards and discharge standards (DOE, 1997b). The standards determined and described in the ECR '97 are being applied nationwide and serve as a national guideline for monitoring. The standards set by the DOE are comprehensive and applicable to all concerned projects including industry and other sectors.

3.8.2 Strengths of the Administrative Arrangements of EIA

a. *Separate environmental agencies (MoEF and the DOE)*: There is a separate Ministry to deal with environmental issues in Bangladesh. There are two Ministers (one full Minister and one State Minister) responsible to lead the Ministry. Within the MoEF, the DOE has the sole responsibility to administer EIA and act as a coordinating "one stop" facility for applicants.

b. *Authority and functions of DOE as mandated by ECA*: Within the MoEF, the DOE is solely authorized to implement the intents of ECA '95. Its rational and legal authority to exercise technical control over project planning process in the public and private sectors can be usefully directed to achieve stated objectives of EIA.

The DOE has the power to declare any area as Environmentally Critical Area when necessary. According to ECA, the DOE also has the statutory power to revise and set environmental quality standards. This strength of legal mandate for DOE to implement EIA is a measure of the level of Bangladesh government's commitment to an effective EIA system.

c. *The Environmental Court*: Environment courts are established under the Environmental Court Act of 2000 to enforce the ECR of 1997. There are six environmental courts in six administrative Divisions in Bangladesh. The Environmental Court Act states the jurisdiction of Environmental Court (Section 5), procedure and power of Environmental Court (Section 8), right of appeal by the aggrieved person, and arrangement of Environmental Appeal Court (Section 12) (DOE, 2000). The existence of the environmental court promises an opportunity to force the proponent to comply with EIA provisions.

d. *Involvement of environmental agencies in the project approval process*: For public projects, a Development Project Proforma (DPP) is submitted to the ECNEC in the Planning Commission, along with a project feasibility study report for the approval of the project. The DPP has a section that requires information about the potential impacts of the project and subsequent mitigation measures to protect the environment from impacts from the project. Obviously, this section forces the proponent to conduct EIAs parallel with project Feasibility Studies.

3.8.3 Weaknesses of the Institutional Arrangements of EIA

a. *Weakness of the legal provisions of EIA*: Experiences in developing countries in Asia and elsewhere suggest that laws of EIA can be conflicting, ambiguous, and inadequate (Mendez and Diaz, 2000, p. 52) for effective EIA practice when they are introduced for the first time. EIA in Bangladesh experiences the same situation. EIA laws (ECA and ECR) were developed more than one decade ago. No comprehensive revision, amendment, or reforms of the laws have been done yet. Therefore, despite their strengths some weaknesses still remain.

b. *Stages of the EIA process are not clearly defined*: While the legislation and policies that underlie EIAs are well founded, the legislation does not spell out EIA procedures clearly and comprehensively. The EIA legislation clearly mentions screening only as one of the stages of the EIA process (ECR of 1997). The legislation does not explicitly define the procedure of other stages, such as scoping, alterative analysis, prediction and evaluation of impacts, and monitoring and auditing.

The interviewees underscored that the ECR does not need to explain the details of methodological and procedural requirements that are pertinent to particular sectoral projects. These requirements can be comprehensively illustrated by the respective sectoral guidelines according to the local needs and nature of the project. However, the generic stages of EIAs and other procedural requirements (the process of review of EIS and community involvement) should be defined by legislation. In the countries, where the EIA system is relatively matured and better performed, procedural activities are clearly and comprehensively stated by EIA rules (for example, the ECR of 2000 in Australia and CEQ NEPA Regulations in the United States).

The ECA or ECR does not particularly mention social impacts and their mitigation. As a densely populated country, people in Bangladesh are frequently affected by development projects and therefore clear and detail rules about social impacts should be there. For example, often cash compensation is considered by the proponents for the displaced people for losing their sources of livelihoods (land or jobs). However, the EIA rules do not cover all losses caused by a project.

c. *Validity of ECC*: According to the ECR '97, the validity of ECC is only for 1 year after it is issued (DOE, 1997a). The current provision of renewing the ECC every year is not suitable for many projects. This controversial provision often raises tension between the DOE and the proponents, such as Water Resource Ministry or the Ministry of Communication. This provision also increases the workload of DOE and causes wastage of time and money of proponents. In India, the validity of ECC for a project is for minimum 5 years and maximum 30 years based on the category of projects. (MoEF, in India, 2000). Therefore, the validity of ECC for some particular projects in Bangladesh should be for more than 1 year.

d. *No provision of EIA for expansion of new and ongoing project*: Expansion of new and ongoing project is a common phenomenon in Bangladesh. An expanded project may cross the threshold limit and therefore may have additional environmental impacts to the originally proposed project. The ECC '95 has not specified whether EIA is required for expansion of new or ongoing project. In the absence of this provision, the proponent may attempt to avoid undertaking EIA study for the additional impacts. To address this issue, Environmental Conservation Amendment Act 2009 has included a provision that requires EIA of project expansion.

e. *No clear provision of public involvement in EIA process*: There is no explicit legal provision for public participation in the preparation or review of EIA report, or during the implementation of mitigation measures. The ECA only includes the provision (Section 8) that any person affected or likely to be affected as a result of pollution or degradation of the environment may apply to the DG of the DOE for remedy of the damage (DOE, 1995, p. 159). Although all EIA guidelines require public participation in all stages of EIA (DOE, 1997b), it is not enough to force the proponent to involve public in the EIA process since it has no legal basis (Ahammed and Harvey, 2004).

f. *Site clearance affects EIA study*: In the ECR '97, under the schedule 7, there is a provision of issuing a site clearance certificate by the DOE for project proponents before issuing the ECC. Once the proponents get site clearance, they are allowed to undertake development works such as land development at the project site. This is a great loophole in the existing EIA rules. In fact, the site clearance allows the proponents to start working at the project site before the approval of EIS and the issuance of ECC. Since the proponent has already invested on the site, it often becomes morally and technically difficult for the DOE to reject the EIS even if it is of poor quality. Rather, the provision of site clearance forces the DOE in many cases to approve the EIA reports hurriedly.

g. *Inadequate judicial control over the EIA system*: According to ECA '95, the parties affected by a development project are not entitled to go to court directly. This requires parties to pursue administrative remedies through the DOE before they are entitled to proceed to judicial process. The Section 17 (2) of ECA '95 states that "no court shall take cognizance of an offense or receive any suit for compensation under this Act except on the written report of an Inspector of the DOE or any other person authorized by the DG of DOE" (DOE, 1995).

The consequence of this provision is that lengthy administrative process may delay actions against the polluters. According to ECR, there is a provision of hearing to be arranged by the DOE. After the hearing, the DG of the DOE assigns an inspector to investigate the case and submit a report. Based on the report if the DG is convinced, then the victim is permitted to go to court. This process may open up the avenue for personal negotiation between the DOE staff and the polluter to settle the issue outside the court and as a result may become a source of corruption.

h. *Appeal by the proponent against the decision of the DOE*: After the review of an EIS, if the DOE rejects it, the proponent can appeal to the Appeal Committee if it is unhappy with DOE's decision. However, the provision can be a barrier to the effective implementation of appeal procedures.

There is no provision for an independent Appeal Committee in Bangladesh. Since there is no independent review committee to review the EIA reports, the Appeal Committee should be truly independent, as suggested by the interviewees.

i. *Lack of adequate EIA guidelines*: In an effective EIA system, adequate sector-wise procedural and technical guidelines for project proponents and EIA consultants are necessary. In the absence of EIA guidelines for dealing with specific sectors of development, the legal adherence may be difficult to achieve (Briffett, 1999).

While there are sectoral and technical EIA guidelines available for proponents and EIA practitioners in many countries, only four public sectors have EIA guidelines in Bangladesh. These sectors are the Industrial Sector, Water Resource Sector, Communication Sector, and LGED Sector. Among the national NGOs, only the BRAC has its own EIA guidelines to implement small-scale projects for rural development. There are no technical guidelines, for example, for the review of EISs or for community participation in the EIA process.

It is also important to produce local EIA guidelines since the guidelines would better reflect local conditions, including the laws, institutions, standards, and procedures for the country in question. Second, the process for creating these guidelines may in itself have a useful effect in promoting information flow, awareness, and interdepartmental cooperation. Third, although the proponents may follow donors' guidelines to conduct EIAs (such as ADB and WB), these guidelines contain generic issues and may not always be appropriate for local audiences and needs.

j. *Age-old environmental quality standards*: Standards are set on the scientific information of environmental conditions, economic situation, and technologies available in a country. When these conditions change, it becomes necessary to revise the environmental standards. This is particularly important in Bangladesh since developmental activities have resulted in excessive environmental degradation. The EQS were set in 1995, and there is no periodic review in response to the changing environmental and economic conditions due to population pressure, rapid urbanization, industrialization, and excessive exploitation of natural resources.

k. *Absence of strategic environmental assessment*: Recent development in the EIA arena is the emphasis on strategic environmental assessment (SEA) or policy-level assessment. That means that rather than only conducting project level impact assessment, it is more advantageous to conduct EIA at the policy development level. This would provide the decision makers with more time to consider environmental consequences at an early stage. SEA would also allow the consideration of cumulative impacts of various projects. At present, there is no legislation of SEA in Bangladesh. The reality is that the country is still in the process of fine-tuning its project level EIA practice and learning from its past errors. The decision makers are rather unfamiliar about the concept of SEA. There is an urgent need for introducing SEA in a country like Bangladesh where major development programs are being implemented by a number of local and international agencies.

3.8.4 Weaknesses of the Administrative Arrangements of EIA

Every EIA system is different, reflecting the political, cultural, and socioeconomic conditions of the country it is prepared for. No single EIA model is appropriate for all countries; this is neither possible nor desirable. An EIA framework or components from one country (or international organization) may not be readily applicable into another without significant adaptation. Like many other developing countries, EIA in Bangladesh is imported from developed countries. It was introduced carefully in the context of

Bangladesh in collaboration with donor agencies keeping in mind the socio-cultural and economic aspects of the country. Despite that, there are some weaknesses in the administrative arrangements of EIA in Bangladesh that need addressing for more effective EIA.

a. *Inadequate budget for DOE*: DOE is under-budgeted with its relatively broad mandate including the implementation of EIA. The revenue budget allocation for the DOE was less than US$0.5 million in 2003-2004. This budget was only 0.0957% of total Annual Development Budget of the GOB. This revenue budget covers only salary of the DOE staff, limited internal travel, and certain basic logistics. There is no budget allocation for activities such as environmental awareness programs, research, and regular monitoring and auditing. The DOE depends on donor funds for these activities that are often conditional and uncertain. Therefore, continuous enforcement and improvement of EIA become difficult.

There are two departments under the MoEF; the Department of Forest (DOF) and the DOE. A recent study of the WB shows that DOE's annual budget (both revenue and donors' grant) is 20% of total budget allocated for DOF though the DOE is mandated to implement the broader part of the activities of MoEF. Moreover, 30% of the budget for the DOF is spent for the payment of salaries compared with 60% of the DOE budget allocated for the same purpose (World Bank, 2006, p. 69). Overall, budget for MoEF is only 0.38% of the total national budget and it is not proportionately distributed between DOE and DOF.

b. *Inadequate manpower and structure of DOE*: While ECA '95 assigns the DOE with broader environmental responsibilities, its manpower is very inadequate to effectively discharge its responsibilities. The volume of tasks has been increasing over the past 18 years, but the manpower has not increased since its establishment in 1993. To date, the DOE has only 244 staff positions of which 101 are managerial or technical. This strength of manpower is lower than that in South-East Asian countries (World Bank, 2006, p. 70) such as Thailand or Vietnam and South Asian countries such as Nepal, Sri Lanka, or Pakistan.

Although the review and approval of EIA reports and the issuance of ECC are the dominant functions of the DOE, the EIA unit of DOE responsible for performing theses activities is highly understaffed. There are only five office bearers (two first-class officers, one second-class officer, and two third-class officers) responsible for EIA report review, approval, and issuance of ECC at the DOE head office. At the Divisional offices, there are only two officers, on an average, responsible for the issuance of ECC. It is to note that these officers also often perform other relevant environmental tasks (such as legal proceedings against illegal development). This situation is going to be more acute with the increase in applications submitted by proponents each year. In 1996, the number of applications for ECC was 421; it increased to 1300 in 2001 and to 2791 in 2006.

The actual time taken to process and issue Site Clearance and ECC for IEE and EIA is always more than the time specified by the ECR 1997. Given the limited number of staff, it is difficult for the DOE to comply with Section 7 which calls ECCs to be issued within 30 days. Furthermore, only a limited number of staff within the DOE is assigned to oversee and monitor the implementation of EIA at post-EIA stage. As a result, responsibility for monitoring is often left to the proponents without the DOE supervision, creating risks that the mitigation measures are not carried out properly and adequately.

Also, the DOE is unable to consistently respond to complaints largely due to the lack of well-positioned field resources. In the absence of offices at District and Upa-Zilla levels, the DOE offices at divisional level have large volume of tasks due to large jurisdiction. The absence of field-level offices significantly limits local awareness about environmental issues and the participation of local key actors.

c. *Expertise of DOE officials on EIA*: In addition to the lack of adequate staff, there is also a lack of staff in the DOE with expert knowledge on EIAs. This affects the quality control of EIA process including the review of the quality of EISs, the design of mitigation measures, and the quality of the monitoring data. In particular, reviewers and decision makers in the DOE must have appropriate training, knowledge, and skills to be able to identify flaws in an EIS and advise the proponents and consultants how to improve its quality.

d. *Frequent change of top officials and lack of leadership*: In the DOE, no permanent leadership has developed as the DG, the head of the DOE, is appointed from outside (particularly from government administrative service) on deputation usually for 3 years. The leadership of DG has been severely compromised by the fact that the person in charge of this position has changed more than once a year (World Bank, 2006, p. 72) in the past 15 years.

e. *Role of private consultants*: Developers or proponents hire consultants to conduct EIA of development projects that they are proposing. Their intention is to get an EIA done that would highlight the benefits and justify the proposal in order to obtain environmental clearance from the DOE or from the donor agencies for the purpose of fund clearance. It is therefore the job of the consultants to satisfy the proponent's requirements rather than carrying out objective EIAs. In addition, there are no codes of conduct by which the activities of the consultants are governed.

f. *Weak and inadequate interagency coordination mechanism*: Coordination is one of the central attributes of regulatory EIA process management. In Bangladesh, there is no formal and permanent coordination mechanism between the DOE/MoEF and the other actors. Weak intersectoral coordination and cooperation at national level between the DOE (and also MoEF) and other sectoral agencies often hinders smooth operation and execution of EIA regime.

3.9 CHAPTER SUMMARY

In this chapter, the legal and administrative arrangements for the EIA system in Bangladesh were reviewed. The review identified both strengths and weaknesses. The review showed that there is a clear legislative foundation for EIAs along with a separate administrative agency that has adequate legal authority. The review also identified the constraints of the current legal and administrative frameworks that are likely to inhibit an effective EIA system in Bangladesh. The study indicates that the EIA legislation (the ECR) does not clearly and comprehensively mention key procedural stages of the EIA process. Some of the EIA requirements mentioned by the legislation are also ambiguous. The DOE, as the EIA implementing agency, is under-budgeted with inadequate staff, unstable leadership, and a lack of expert human resources.

Although the institutional arrangements of EIAs in Bangladesh may give a satisfactory impression of the EIA system having a good foundation, the institutional arrangements are insufficient to represent the system if the methodological requirements (the stages of the EIA process) of the EIA process are not fully addressed in practice. The real test of an effective EIA system also lies in the implementation of the methodological requirements of the EIA process. With this in mind, Chapter 4 discusses the quality of EISs in Bangladesh.

REFERENCES

Abaza H, Bisset R, Sadler B. Environmental impact assessment and strategic environmental assessment: towards an integral approach. Geneva, Switzerland: United Nations Environmental Program; 2004.

Ahammed AKMR, Harvey N. Evaluation of environmental impact assessment procedures and practice in Bangladesh. Impact Assess Project Appraisal 2004;22(1):63–78.

Briffett C. Environmental impact assessment in East Asia. In: Petts J, editor. Handbook of environmental impact assessment: environmental impact assessment in practice: impact and limitations, vol. 2. Oxford: Blackwell; 1999. p. 143–67.

Department of Environment (DOE). The Environmental Conservation Act. Dhaka, Bangladesh: Ministry of Environment and Forest; 1995.

Department of Environment (DOE). The Environmental Conservation Rules. Dhaka, Bangladesh: Ministry of Environment and Forest; 1997a.

Department of Environment (DOE). EIA Guidelines for Industries. Dhaka: Ministry of Environment and Forest, Government of Bangladesh; 1997b.

Department of Environment (DOE). Environmental Court Act. Dhaka: Ministry of Environment and Forest, Government of Bangladesh; 2000.

Goodland RJ, Edmondson V, editors. Environmental assessment and development. Washington, DC: World Bank; 1994. p. 29–34.

Government of Bangladesh. The Constitution of the People's Republic of Bangladesh. Dhaka: Ministry of Law, Justice and Parliamentary affairs; 1973.

McHarg IL. Design with nature. Garden City, NY: Natural History Press; 1969.

Mendez JM, Diaz C. Environment in transition: Cambodia, Lao PDR, Thailand and Viet Nam. Manila, Philippines: Asian Development Bank; 2000.

Ministry of Environment and Forest (in India). Environmental protection rules of 1986, Government of India; 2000. 1986, viewed 2 May 2010. http://moef.nic.in/modules/rules-and-regulation

Ministry of Environment and Forest (MoEF). National Environmental Policy. Dhaka: Government of Bangladesh; 1992.

Modak P, Biswas AK. Conducting environmental impact assessment for developing countries. New York: United Nations University Press; 1999.

World Bank. Bangladesh: country environmental analysis. Dhaka, Bangladesh: World Bank Office; 2006.

CHAPTER 4

The Quality of Environmental Impact Statements

Contents

4.1 INTRODUCTION

The effectiveness of the environmental impact assessment (EIA) system of a country depends on many aspects, but among these, the quality of the environmental impact statements (EISs) is of particular importance. An EIS is the fundamental indicator of an effective EIA system for the reason that the information presented in the report reflects the technical and scientific quality of the EIA process. The EIS document is the only way to incorporate and present scientific knowledge in an EIA study.

An EIS is the product of an EIA process, and as such, the quality of the document is likely to be closely associated with the quality of the whole EIA process. This can be supported by the findings of the empirical studies of Wende (2002). In her analysis of the impact of an EIS on the effectiveness of EIA, Wende (2002) shows that there is a clear relationship between the quality of an EIS and the effective EIA process.

4.2 REVIEW PACKAGE FOR ASSESSING THE QUALITY OF EISs

In this study, the quality of EISs was reviewed following the review procedure developed by Lee and Colley (1992). The Lee and Colley (1992) method has been widely used by researchers (for example, Badr et al., 2011; Barker and Wood, 1999; Canelas et al., 2005; Cashmore et al., 2002; Hughes and Wood, 1996; Pinho et al., 2007; Sandham and Pretorius, 2008) as shown in Table 4.1.

However, the quality of EISs has to be assessed taking into account the regulatory and procedural context in which they are prepared. Lee and Colley (1992) developed review criteria in the context of the United Kingdom. Therefore, studies held elsewhere (other than the UK context) have taken this into account. For example, Sandham and Pretorius (2008) and Badr et al. (2011) adapted the review criteria such that they would be suitable to the South African and Egyptian context, respectively. With this in mind, some criteria (* marked) were added to the Lee and Colley (1992) review criteria to make the review package suitable to the context of Bangladesh (Table 4.2).

4.3 DESCRIPTION OF REVIEW CRITERIA

The criteria are divided into three categories, namely, area, category, and subcategory. At the top, there are four areas according to Lee and Colley (1992):

Table 4.1 Methods used by authors to assess EIS quality

Authors	Assessment methods and sectors	Grading procedure
Sandham and Pretorius (2008)	Reviewed the quality of 28 EISs of different sectors in South Africa using review criteria	Used alphabetic symbols to grade each criterion, where A = well performed, B = generally satisfactory, C = just satisfactory, D = just unsatisfactory, E = poorly attempted, F = did not attempt
Pinho et al. (2007)	Assessed quality of EISs of small hydropower projects in Portugal using review criteria	Used numeric values, where Very poor = 0.0-0.4, poor = 0.5-0.9, fair = 1.0-1.4, fair/good = 1.5-1.9, good = 2.0-2.4, very good = 2.5-3.0
Canelas et al. (2005)	Assessed and compared the quality of EISs of different sectors between Portugal and Spain	Used alphabetic symbols, where A = excellent, B = good, C = satisfactory, D = poor, E = very poor
Cashmore et al. (2002)	Assessed quality of EISs of different sectors in Greece	Used alphabetic symbols, where A = well performed, B = generally satisfactory, C = satisfactory, D = just unsatisfactory, E = not satisfactory, F = very unsatisfactory
Barker and Wood (1999)	Assessed and compared the quality of EISs of different sectors in eight EU countries	Used alphabetic symbols, where A = well performed, B = generally satisfactory, C = satisfactory, D = just unsatisfactory, E = not satisfactory, F = very unsatisfactory
Hickie and Wade (1998)	Used 19 review questions to review the quality of 14 EISs of different sectors in United Kingdom	Used numerical weakness index, where 6 = very poor, 5 = poor, 4 = unsatisfactory, 3 = satisfactory, 2 = good, 1 = excellent
Lee and Colley (1992)	Used review criteria developed in the context of United Kingdom	Used alphabetic symbols, where A = well performed, B = generally satisfactory, C = satisfactory, D = just unsatisfactory, E = not satisfactory, F = very unsatisfactory

Source: Compiled by the authors.

Area 1: The description of development and baseline conditions,
Area 2: Identification and evaluation of key impacts,
Area 3: Environmental management plan and follow-up, and
Area 4: Presentation of EIS.

Table 4.2 Criteria for the review of EIS quality in the context of Bangladesh
1. Description of the development and baseline
    1.1. *Description of development*
        1.1.1. Background and objectives of project
        1.1.2. Design and size of the project
        1.1.3. Some indication of physicalpresences
        1.1.4. Nature of production process
        1.1.5. Nature and quantities of raw materials
        1.1.6. *Policy and legal framework for EIA
        1.1.7. *EIA aim and objectives
        1.1.8. *Limitation of study
    1.2. *Description of site*
        1.2.1. Land area taken by the development site
        1.2.2. The use of land taken
        1.2.3. Estimated duration of project alternatives
        1.2.4. Number of workers and means of transport
        1.2.5. Means of transporting raw materials
    1.3. *Waste generation*
        1.3.1. Types and quantities
        1.3.2. Production process and waste generation
        1.3.3. Treatment of wastes generated
    1.4. *Environmental description*
        1.4.1. Environment expected to be affected
        1.4.2. Offsite impacts
    1.5. *Environmental baseline*
        1.5.1. Description of important components
        1.5.2. *Natural physical environment
        1.5.3. *Biological environment
        1.5.4. *Socioeconomic environment
        1.5.5. Methods and sources of data with justification
        1.5.6. Future state of environment without project
2. Identification and evaluation of key impacts
    2.1. *Definition of impacts*
        2.1.1. Impact types
        2.1.2. Impacts with regard to human, ecology, etc.
        2.1.3. Impacts of accidents
        2.1.4. Impacts as the deviation of baseline
    2.2. *Identification of impacts*
        2.2.1. Methods used for impacts
        2.2.2. Justification of methods used
    2.3. *Scoping*
        2.3.1. Arrangements for scoping
        2.3.2. Methods of collection of opinions
        2.3.3. Inputs from stakeholders
        2.3.4. Selection of key impacts to be investigated

Table 4.2 Criteria for the review of EIS quality in the context of Bangladesh—cont'd
    2.4. *Prediction of impacts*
        2.4.1. Data gaps and uncertainty
        2.4.2. Methods used with justification
        2.4.3. Quantitative expression where possible
        2.4.4. *Uncertainty
    2.5. *Assessment of impact significance*
        2.5.1. Significance of impact on community and environment
        2.5.2. Methods used for evaluation of impacts
        2.5.3. Justification of methods used
    2.6. *Community involvement*
        2.6.1. Description of community affected
        2.6.2. Involvement of community
        2.6.3. Methods of community involvement
        2.6.4. Inputs from community
3. Alternatives and environmental mitigation
    3.1. *Alternatives*
        3.1.1. Alternative sites
        3.1.2. Alternative process, design, and activities
        3.1.3. Selection of alternatives
    3.2. *Scope and effectiveness of mitigation measures*
        3.2.1. Description of adverse impacts to be mitigated
        3.2.2. Mitigation measures with justification
        3.2.3. Residual impacts
    3.3. *Commitment to mitigation*
        3.3.1. Implementation arrangements
        3.3.2. Monitoring program
        3.3.3. *Parameters to be monitored
        3.3.4. *Feedback and reporting mechanism
4. Presentation of EIS and communication
    4.1. *Layout*
        4.1.1. Introduction and brief description
        4.1.2. Logical arrangements of information
        4.1.3. List of references
    4.2. *Presentation*
        4.2.1. Comprehensible to nonspecialist
        4.2.2. Defining technical terms
        4.2.3. Presented as an integrated whole
    4.3. *Emphasis*
        4.3.1. Potentially severe adverse impacts
        4.3.2. Unbiased statements
    4.4. *Executive summary*
        4.4.1. Findings presented in a nontechnical way
        4.4.2. Recommendations

Source: Kabir and Momtaz (2012, p. 99).

Under each area, there are categories (for example, 1.1, 1.2, ..., 2.1, 2.2), and under each category, there are subcategories (for example, 1.1.1, 1.1.2, ..., 2.1.1, 2.1.2). All the areas, categories, and subcategories are referred to as tasks.

4.4 SELECTION OF EISs

A total of 30 EISs of different projects (Appendix 4.1) were selected irrespective of different development sectors and the year of the EIS preparation. This study covers the EISs prepared between 1995 and 2008. There have been 20 full EIAs for major projects conducted in Bangladesh on average per year (DOE, 2009). With this in mind, it can be estimated that, after the initiation of formal EIAs in 1995, the total number of EISs subjected to full EIAs is about 280. Therefore, the sampled 30 EISs for this study are representative of the EISs prepared in Bangladesh and allow a general understanding of the quality of EISs in Bangladesh. A list of the EISs, along with respective sectors, is given here (Table 4.3).

4.5 ASSESSMENT PROCEDURES

Figure 4.1 shows a schematic view of the assessment procedure, based on Lee et al. (1999), starting with the subcriteria and finishing with the overall assessment of EISs. Each subcriterion was graded on the basis of the adequacy and quality of the materials provided. The judgment of the adequacy of

Table 4.3 Number of EISs by sector

S. no.	Name of sectors	Number of EISs	Major characteristics
1	Water Resource Management	9	All are public projects excepting two that are funded by donor agencies. Two are funded by government only
2	Industry	5	Three projects are private projects and two projects are donor-funded public projects
3	Energy, Power, and Mineral Resources	11	All projects are public-funded projects and private (owned by multinational companies) projects
4	Infrastructure (roads, bridges, and urban projects)	5	All projects are public projects and funded by donor agencies

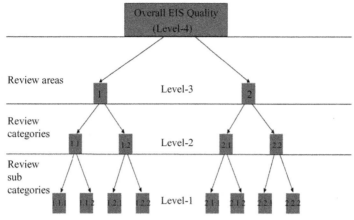

Figure 4.1 Schematic view of assessment procedure from subcategory to the overall assessment of quality. (For color version of this figure, the reader is referred to the online version of this chapter.) *Source: Compiled by the authors.*

information in each criterion and subsequent gradation was done based on the following issues:

- what it (information) must contain;
- what it (information) could contain; and
- what it (information) could reasonably be expected to contain (Weston, 2000).

The review commenced at the lowest level (Figure 4.1), that is, the subcriteria level (Level 1). Each subcriterion was awarded an alphabetic symbol as a grade according to the quality of information presented under that subcriterion. An average grade was then calculated for each respective category at Level 2. In this way, the average grade was calculated for each area at Level 3. Finally, from the grades given to each area, it was possible to arrive at an overall average grade of the EIS (Level 4).

Alphabetic symbols were used to grade the subcriteria, criteria, and areas. Table 4.4 shows how the alphabetic symbols correspond with the grades. There are six alphabet symbols (A, B, C, D, E, and F) used to grade the criteria. Where a subcriterion or a criterion was not relevant, "NA" was placed in the record. Attributing grades for each criterion forces the reviewers to identify and clarify, overall, what information is missing and what is the quality of existing information in terms of its clarity, organization, and presentation. Finally, using such a review process, the reviewers can justify any deficiency with confidence and put requests for further information into practice.

Table 4.4 Alphabetic symbols and the definition of grades

Alphabetic symbols	Grades	Basis of grading: what does it mean
A	Excellent	Well performed. Tasks are fully completed, all relevant information provided accurately and adequately where necessary
B	Good	Generally satisfactory. A completed EIS with only minor omissions and inadequacies
C	Just satisfactory	Parts of the EIS are well attempted, but, on the whole, satisfactory only because of some major omissions and the limited coverage of information
D	Poor/just unsatisfactory	Revealing significant omissions and inadequacies. Very limited coverage
E	Very poor/ unsatisfactory	Revealing significant omissions and inadequacies. Important tasks are poorly attempted
F	Fail/very unsatisfactory	Not attempted at all
N/A	–	Not relevant or not applicable

A collation sheet (Appendix 4.2) was used to record the grades against each subcriteria and criteria and subsequent area of an EIS. Moreover, the principal strengths and weaknesses of the EISs were recorded in brief. Furthermore, the length the EISs (number of pages), the size of EIA team, the sources of funding (government, private, or donors), and types of proponents (government or private) were recorded while reviewing the EISs.

Weston (2000) states that the attributing of an individual grade-point (be it numeric or nonnumeric) to an individual criterion is inherently subjective. In order to overcome this limitation, Lee et al. (1999) suggest that each EIS should be reviewed by two reviewers. In this study, the EISs were reviewed independently by one of the authors and one EIA expert. Both reviewers were quite familiar with knowledge and skills concerning EIA issues. Where any difference was found in symbols for a criterion, then that criterion was carefully reviewed for the third time. The final result was then accepted, balancing the findings obtained from two reviewers. Appendix 4.3 shows the summary results of the 30 EISs reviewed.

4.6 OVERALL QUALITY OF EISs IN BANGLADESH

Figure 4.2 shows the overall quality of EISs of 30 sampled projects in Bangladesh. From the graph, it can be seen that 66% of EISs are graded

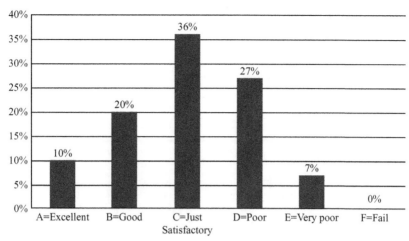

Figure 4.2 Overall quality of EISs in Bangladesh. *Source: Kabir and Momtaz (2012).*

excellent to just satisfactory (A–C) and 34% EISs are graded poor to very poor (D–E). No EIS is found with an F or fail grade.

The findings presented in Figure 4.2 reveal that the overall quality of EISs in Bangladesh is "satisfactory," in general (where 66% of EISs are just satisfactory and above). These findings broadly correspond to the findings of other similar studies (for example, Badr et al., 2011; Barker and Wood, 1999; Cashmore et al., 2002; Glasson et al., 1997; Sandham and Pretorius, 2008) where they have found that the overall quality of the EISs of the particular country under examination is generally satisfactory.

4.7 QUALITY OF EISs BY AREA

In addition to the overall quality of EISs, more details are required to identify the strengths and weaknesses of the EISs. Therefore, the researchers have analyzed the quality of the EISs by area. Figure 4.3 shows the results of the quality of sampled EISs in four areas. The results of the analysis are discussed below under the headings of the four main review areas. This analysis is important as it helps to identify the tasks that need further attention for future improvement.

4.7.1 Area#1: Description of Development and Baseline Conditions

Area#1 is one of the two better performed areas in the EISs where 83% of EISs are satisfactory. The relatively straightforward and readily available tasks

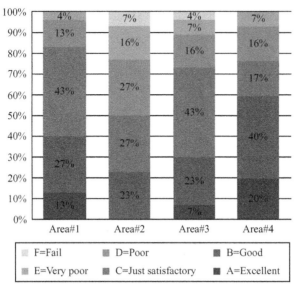

Figure 4.3 Quality of the EISs by area. (For color version of this figure, the reader is referred to the online version of this chapter.) *Source: Prepared by the authors based on quality review information.*

tend to be performed best. These include the background of the project, the EIA aims and scope, the policy and legal framework for EIA, and the description of the project. Previous studies, such as Lee et al. (1999), Barker and Wood (1999), and Cashmore et al. (2002), have found that such descriptive tasks tend to be of higher quality. However, there are deficiencies in some tasks, such as in the baseline information and the discussion on limitations of the EIA study.

4.7.1.1 Baseline Data
The collection of baseline data on relevant project locations is the most important basis for conducting an EIA study. In this study, 100% of the EISs contain baseline data, but the coverage and the quality of baseline information tend to be poor. Data indicate that about 30% (9) of EISs are unsatisfactory (Figure 4.3). In most cases, collected baseline data are found to be voluminous, all inclusive and descriptive rather than precise. Therefore, it is often difficult for a reviewer to find out which data are exclusively relevant to the potential environmental impacts of the proposed project. This is a significant weakness given that baseline data form the foundation of prediction and evaluation of impacts.

In many cases, less important data have been collected, but relevant and important data in relation to the nature and type of the project are missing. For example, in order to consider the potential impact of a gas pipe line project or power transmission line project, it is necessary to collect detailed information on physical or ecological characteristics, such as the deterioration of soil quality and the disturbance of wildlife by the projects. However, the EISs of these types of projects are often found to be deficient of such important information. Similarly, in the case of socioeconomic data, much of the data collected are often not relevant and, therefore, not useful for impact prediction. The reason behind the collection of all-inclusive, descriptive, irrelevant, and inconsistent data may include the absence of or poorly performed scoping, and the lack of professional knowledge of the data collectors and consultants of what information is required for a particular impact prediction process.

The baseline information on ecological and physical conditions should be quantitative. This is important, as the accuracy and plausibility of much of the remainder of EISs often depend on quantitative data. There is a lack of quantifiable baseline biophysical and ecological data in many of the sampled EISs. For example, the EIS of Meghnaghat Power Plant Project states that "the rate of fish production in the river nearby the project is decreasing" (GOB, 2000) without any statement about how much.

Many of the EISs in this study have claimed that they have collected primary data as well as secondary data in order to fill the data gaps. However, most of the EISs gave few details about the methods used for the collection of primary data. The majority of the EISs in this study stated that the data were collected by "a multidisciplinary team of experts" but did not provide their academic background and expertise.

4.7.1.2 The Limitations of the EIA Study

Surprisingly, most EISs did not provide description of any limitations that may be encountered by the consultants in obtaining details at different stages of an EIA study. This finding can be substantiated by other similar studies, such as by Gray and Edward-Jones (2003). Information on the limitations of studies has great value for future practitioners. Those that did mention limitations mostly mentioned time constraint.

4.7.2 Area#2: Identification and Evaluation of Key Impacts

This is the most important area of an EIS as it contains the results of an EIA study. However, this is also the review area with the lowest performance rating where 50% (15 EISs) of EISs are assessed as being "poor" and/or "very poor" (Figure 4.3). Common deficiencies include limited information about scoping, failure to provide adequate explanations of methods or techniques used to predict and evaluate the impacts, inadequate coverage, quantification of impacts, poor evaluation of impact significance, and failure to involve the potentially affected community's input in the assessment of impacts.

4.7.2.1 Scoping

Scoping is an important phase in conducting EIAs whereby the geographic regions and issues to be considered in an EIA are determined (Momtaz, 2002, p. 175). The scoping exercise is intended to confirm the impacts that are expected to be significant and to determine the time periods and locations over which they must be studied. Scoping is found to be either absent or poorly described in the reviewed EISs, although it is widely recognized that scoping can be a catalyst for improvement.

In this study, 50% of the sampled EISs are unsatisfactory in relation to the information about scoping. In general, EISs make only brief and poor mention of the scoping process. For example, two EISs from the Water Resource Development Sector and the Infrastructure Development Sector projects state merely that "a scoping process was undertaken to select Important Environmental Components (IECs) and Important Social Components (ISCs)" (GOB, 2001, p. 22; GOB, 2007, p. 3). The EISs did not mention the scoping process in detail including the methods of scoping, the involvement of stakeholders, the development of Terms of Reference (TOR), data types and sources, and the spatial (and temporal) distribution of potential impacts. These deficiencies are found consistently in the EISs, despite the fact that the absence of scoping or poorly done scoping may result in inappropriate, time consuming, and costly EIA studies.

The involvement of a broad range of stakeholders, including both experts and nontechnical persons, is widely accepted to be an essential element of an effective scoping process. However, there is a lack of detailed descriptions about the stakeholders who have been included in the scoping process. Where involvement by stakeholders has taken place, most EISs simply contain a list of stakeholders who have been contacted.

4.7.2.2 Identification of Impacts

All of the EISs examined in this study identify the potential impacts of the development projects, but none do it with full coverage. All of the EISs identify social and biophysical impacts, but most of them fail to identify health impacts on the affected people, in general, and the vulnerable groups, such as children, pregnant women, and the poor, in particular. In Bangladesh, where people are frequently affected by development projects, this issue should receive serious consideration by EIA consultants. This omission is not unique to developing countries. Results from Steinmann's (2008, p. 632) study in the United States show that more than 62% of EISs do not mention health impacts.

Similarly, hardly any EIS makes explicit mention of cumulative impacts (CIs). Bangladesh is a land-hungry country, and therefore, projects are often located closely to each other. The Industrial EIA Guidelines of DOE suggest consideration of CIs during an EIA study (DOE, 1997). However, few EISs even mention the term. This may be due to minimal attention being paid to the definition of spatial and temporal boundaries for the study and to the delineation of other existing projects in the same environs.

4.7.2.3 Prediction of Impacts

Once the impacts are identified, a range of methods are employed to predict impacts. The use of a number of methods increases the credibility of impact results derived from an EIA (Momtaz, 2002, p. 175). The methods may range from a simple checklist to complex mathematical models. All of the EISs in this study have used either the checklist or the matrix method to predict impacts. These methods are popular and used widely in all of the jurisdictions where EIAs are practiced. However, despite their popularity, these methods also have limitations. It is not possible to identify indirect effects using these methods, and these methods can be used subjectively. However, hardly any of the EISs reviewed make any reference to these limitations.

Some of the EISs have used mathematical and physical models, in addition to the methods mentioned above, to predict physical and ecological impacts. In this study, however, the EISs make little mention of any justification for using these techniques.

From the EISs reviewed, it can be seen that "expert judgment" is the dominant method used by the consultants for the prediction of impacts. Expert judgment is quick and cost-effective and, therefore, popular in developing countries. Where impacts are not quantifiable, such as social or cultural impacts, expert judgment is unavoidable. However, expert judgment becomes more accurate when the EIA experts have a thorough understanding of the scientific process involved, familiarity with the particular characteristics of the receiving environment and extensive relevant practical knowledge and skills that are involved in the EIA process. Therefore, the reason behind the use of professional judgment should always be justified in the EISs. Most EISs simply state that the "prediction of impacts is the outcome of expert judgment." The process of reaching a point using expert judgment about the prediction of impacts should be clearly stated to make the findings credible and acceptable. Few EISs in this review have a complete and clear record of expert judgment in the prediction of impacts.

Where it is possible, the predicted impacts should be stated quantitatively. Without quantification, the extent of the magnitude of impacts cannot be properly explained or made credible. It is difficult to quantify social and cultural impacts, whereas physical and ecological impacts are easy to quantify. However, this study has found that in many EISs, a number of impacts are not quantified, though this would have been possible. For example, the EIS of a road project (JMBA, 2001) simply states that habitat would be lost but does not provide any precise description in quantitative form.

4.7.2.4 Lack of Adequate Assessment of Social Impacts

A lack of effort in identifying and predicting social impacts in detail is observed in the EISs reviewed. The importance of the social impact assessment is given strong recognition in the EIA guidelines of donor agencies (Asian Development Bank, 2003; World Bank, 1999). Also, one of the requirements of the DOE is a description, with due care, of social and cultural impacts in the EISs. As the DOE states, "It is necessary that social and cultural impacts are given the prominence they deserve in describing the changes expected to result from major development projects" (DOE, 1997, p. 26). The analysis and presentation of social impacts may provide an impetus for achieving the same level and quality of assessment for such impacts as is given to biophysical effects (Lockie et al., 1999; Momtaz, 2003, 2005, p. 41). The assessment of social impacts is important in a densely populated country (983 km^2) like Bangladesh, where project interventions frequently cause problems for people.

While considering social impacts, all of the EISs focus on the displacement of people, loss of houses, lands and other sources of incomes (for example, loss of commercial fishery), and changes in status and employment. These impacts are identified and predicted, but with the simple assumption that all people irrespective of their status and income will be equally affected.

There are vulnerable members in the community, such as women, female heads of households, the ultra-poor, children, indigenous people, and aged persons. Their experience of the impacts will not be the same as the adults or the rich in the community. Resilience of vulnerable groups is far less than other groups in the society. In fact, EISs have collected only minimal socioeconomic information concerning these vulnerable groups and, therefore, fail to predict the differential effect of impacts on them. Thus, the scope and content of social impacts in the EISs are only narrowly defined.

4.7.2.5 Assessment of Impact Significance

The use of the words "significance" or "significant" is found in the EISs under this study without any definition of the criteria used for determining the significant impacts of a project. When used, all too often, the meaning of significance is found to be vague and ambiguous. Any exercise in judging the significance of predicted impacts in the EIS must be undertaken in a clear and transparent fashion. The significance of predicted impacts is often judged on the basis of environmental laws or environmental quality standards set by the environmental agency, social perspective, and public opinions. Few EISs in this study interpret the basis of the determination of significant impacts. Surprisingly, many EISs in this study fail to recognize that impact prediction and the assessment of significant impacts are two different stages.

The simplest way of determining the significance of an impact is to compare its predicted value/magnitude with the national environmental quality standard. There are many qualitative or descriptive environmental and social impacts. Thus the determination of significant impacts is not always possible based on verifiable evidences. Therefore, other criteria such as public opinions and expert judgments are necessary. In that case, the evaluation of the significance of impacts tends to have a subjective dimension, arising from the integration of the values, experiences, and knowledge of the different actors that perform the evaluation. Although subjectivity can never be eliminated, the results of an evaluation of impacts may become more credible if they are obtained by the application of *a priori*-defined methodology with clearly stated assessment criteria. The EISs in this review scarcely define "significant impact" or provide any clear descriptions as to why an impact is significant or insignificant.

4.7.2.6 Community Involvement

The members of a community are a source of valuable local knowledge, and they should be involved in EIA process with due importance being attached to their inputs. It is mentioned in EIA guidelines (DOE, 1997) that communities should be involved in all stages of an EIA, but the review results indicated that community involvement is one of the most neglected processes. Only 53% of the EISs provide adequate information on community involvement including methods applied for community involvement, the timing of community involvement, and concerns of potentially affected communities. Some 43% of the EISs make little mention of the community involvement in the EIA process and another 4% make no mention at all of any public involvement.

The EISs reviewed in this study showed that tools used for community involvement in the EIA process include Focus Group Discussions, workshops, key informant interviews, and public meetings. However, there is little explanation as to why the tools are used and how effectively they are used. The review results indicate that the community is consulted in order to identify environmental impacts. However, there are no clear details about how the community might be involved in other stages of the EIA process, such as in the implementation of mitigation measures and the monitoring process. Some of the EISs only state that the "communities were involved during the EIA study" but do not explain the nature of their inputs and how the inputs have been incorporated into the EIA reports. The deficiency in community involvement in the EIA process was also identified in other studies conducted in Bangladesh, for example, by Ahammed and Harvey (2004), Momtaz (2005), and elsewhere by Formby (1991) and Stolp et al. (2002).

4.7.2.7 Uncertainty

Finally, it has also been observed that a small number of EISs mention uncertainty associated with impact prediction and the significance of impacts. Uncertainty in impact prediction arises from a number of factors: a lack of accurate data, a lack of understanding of the behavior of complex systems, and the imprecise assumptions taken for the application of models to predict impacts (Figure 4.3). All sources of uncertainties should be acknowledged in the EISs. Information about uncertainties makes the predicted impacts credible to decision makers.

4.7.3 Area#3: Mitigation and Monitoring

All of the sampled EISs contain designed mitigation and monitoring plans within their Environmental Management Plans (EMPs) in order to address the significant environmental impacts of the projects. However, the data indicates that a considerable number of EISs (27%) are still assessed as being unsatisfactory in this area (Figure 4.3). Common inadequacies include lack of analysis of alternatives, poor descriptions of adverse impacts, a lack of justifications for the success of mitigation measures, a lack of commitment in public involvement during the implementation of mitigation measures, effective mitigation for social impacts, the costs of mitigation measures, and residual impacts. The consideration of mitigation measures and monitoring is also found to be one of the more problematic areas in other studies, for example, Sandham and Pretorius (2008).

A lack of consistency has been observed between the significant predicted impacts and mitigation measures designed against those impacts. In the reviewed EISs, it can be observed that some of the impacts are not identified as being significant, but mitigation measures are designed for those impacts. On the other hand, mitigation measures are not taken into account for some significant impacts. The omission of mitigation measures for some significant impacts, therefore, has led to poor design of mitigation.

4.7.3.1 Analysis of Alternative

The consideration of alternative options is often neglected by consultants and, therefore, inadequately addressed in EISs. In practice, the site of the project is often selected prior to conducting the EIA. The analysis of alternative options among sites, designs, technology, or input materials provides a major basis for decision makers. In this study, 6% of EISs do not focus on this category at all, and 24% only mention alternative analyses without any detailed description. Only 48% of the EISs have adequate analysis of alternative courses of actions, but most of them are limited to a "no action" alternative. Only 30% of the EISs contain descriptions of all possible options for alternative project designs (Figure 4.3). For example, the KJDRP (Khulna-Jessore Drainage Rehabilitation Project) has analyzed alternative design options using multicriteria analysis (GOB, 1998).

In many cases in Bangladesh, project sites, such as roads and high ways, bridges, and industry, are selected politically rather than on environmental grounds. Therefore, the assessment of alternative sites often becomes a formality. Consideration of alternative locations from an environmental perspective, however, is essential in project planning and minimizing environmental impacts of projects.

In addition to the evaluation of alternative sites for project, EIA studies should focus on the analysis of best alternative technologies and input materials for production process to avoid potential environmental impacts of a project. However, the EISs in this research have not taken these alternative options into consideration.

4.7.3.2 Mitigation Measures for Social Impact Assessment

In this study, the mitigation measures designed for social impacts are found to be inadequate. Displacement of people as a result of development projects is a common scenario in Bangladesh, in addition to the loss of agricultural land and other sources of livelihood.

While the EISs include mitigation measures to address socioeconomic impacts in terms of cash compensation or resettlement with housing and other facilities, little consideration has been given to the mitigation of kinship, social capital, and sources of income that are lost by the resettled people. These impacts may cause impoverishment risks, particularly for the poor and the vulnerable. Only a few EISs in this study have mentioned these issues and designed appropriate mitigation measures.

Mitigation measures for each significant impact (ecological, physical, or social) need to be clearly outlined with the justification for its likely success. This helps decision makers to understand that the action to be taken will be successful in addressing the environmental impacts. However, there is a lack of detailed information in many of the EISs in this study, some mitigation measures are described in detail, and some are not. For example, in addressing ecological impacts, many EISs merely state that ecological damage or loss will be offset or replaced. However, no detailed information has been supplied as to how and to what extent the impacts will and can be mitigated. There is little information about the techniques to be used to restore the ecological damage, estimated time to restore or replace the damage, and the costs of achieving an acceptable degree of mitigation. Therefore, it is difficult for the decision makers to conclude whether the mitigation measures designed are feasible in addressing the respective predicted impacts.

4.7.3.3 Monitoring Program

All the EISs in this review designed monitoring programs within the EMPs, but many of the EISs poorly described the monitoring activity. Without an appropriate monitoring program, it will be difficult to build up the knowledge and experience that is required to improve the effectiveness of mitigation measures and the EIA system at large.

Of the EISs reviewed, few adequately prescribe the ecological, social, and physical parameters that can be monitored during and after the implementation of mitigation measures. Furthermore, 40% of the sampled EISs do not clearly specify the roles and responsibilities of environmental agency, donor agency, community, and the proponent in implementing mitigation actions.

4.7.4 Area#4: Communication and Presentation of an EIS

The communication and presentation of an EIS is one of the two better performed review areas. In this case, more than 77% of the EISs were graded satisfactory, in general (Figure 4.3). Many previous studies, such as Cashmore et al. (2002) and Sandham and Pretorius (2008), have found that tasks related to general structure, layout, and presentation of an EIS tend to be performed well. In this study, all of the EISs include executive summaries or nontechnical summaries (NTSs). Definitions of technical terms are given, thus aiding better comprehension. Most of the EISs use maps, pictures, tables, and charts that make the report well organized and understandable. Also the EISs are found, on average, to be 200 pages in length. This creates a positive impression for the reviewers when reading the EISs.

Despite the glossy paper-bound reports, there are still some deficiencies in this area of EIA reports. When undertaking this study, it was observed that neither the EISs nor the NTSs are understandable to all nonspecialists. One important reason for this is that the reports are not written in Bengali, the language of the people of Bangladesh. EISs in Bangladesh are only ever written in English; therefore, the reports are neither easily understandable nor accessible by all stakeholders, particularly the affected rural community. It must be noted that, as Bangladesh is a developing country, not all of the inhabitants have the same level of education and knowledge.

4.7.4.1 Recommendations by the EISs
All of the EISs reviewed in this study incorporated recommendations under executive summary. However, recommendations mentioned in EISs are often found to be inadequate and vague. No clear direction is available in a "what to do" format from the recommendations as presented in the EISs.

4.7.4.2 Understandability
Some EISs are difficult to read as they contain volumes of data. As mentioned earlier in this chapter, some EISs include irrelevant baseline data and appendices with a minimum of useful information. One EIS for a public project that was reviewed as a part of this study is 375 pages in length of which 315 pages are appendices (GOB, 2006). Another problem is

inconsistent flow of information. For example, in some cases, identified impacts have not been considered in the prediction and evaluation of significance. There is no explanation why the predicted impacts are not considered. Some EISs have inadequate information and are only 60 or 70 pages in length (GOB, 2007), with some of the chapters being only a couple of pages long (GOB, 2006). These "structural errors" call for more coordination among EIA team members, as well as the EISs being professionally designed and carefully edited in order to enhance their clarity.

Although many of the readers of EISs are specialists, many other important readers are not. The latter include, of course, the affected community and other stakeholders, such as NGOs. However, the consultants often target the audiences of government officials, donor agencies, and academics and, accordingly, often prepare EISs using technical jargon and mathematical terms. Thus, the quality and effectiveness of EISs are undermined.

NTSs included in EISs are useful for readers to understanding the impacts of projects quickly. In this study, it is found that 30% of the sampled EISs have poor executive summaries or NTSs. Some NTSs are full of jargons, insufficient and inconsistent. Although not everyone needs to understand everything written in a particular EIS, some basic level of understating is necessary. An empirical study by Sullivan et al. (1996) in the United States reveals that 70% of EISs were not comprehensible to the citizens. If the citizens do not understand an EIS, the consequence is that proponents may lose the opportunity to gain credibility among the affected people.

In addition to an NTS, the EISs must also to be written using simple language that includes minimal technical jargons. As stated earlier, there is no practice of publishing EISs in Bengali, the official language of Bangladesh in order to make EISs accessible to the public. At a minimum, the technical summary should be published in Bengali so as to facilitate basic understanding about the potential impact of a project.

4.8 FACTORS INFLUENCING THE QUALITY OF EISs

Despite the fact that the average quality of EISs in Bangladesh is "just satisfactory," a significant number of EISs (34%) have been found to be "poor" and below poor in grading. In addition, the review results of the quality of EISs by area show that there are deficiencies in information in each area and particularly in the areas of impact prediction, assessment, and mitigation. The major categories of deficiencies include baseline data,

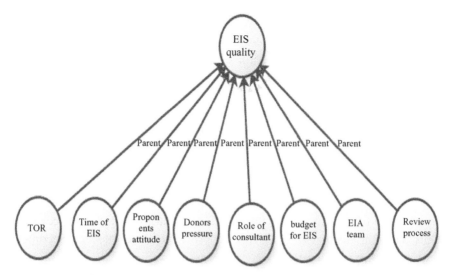

Figure 4.4 Factors influencing the quality of EISs. (For color version of this figure, the reader is referred to the online version of this chapter.) *Source: NVivo.08 data analysis by the authors NB. The word "parent" in figure indicates that each of the factors (child node) such as "EIA team" is related to the EIS quality (parent node).*

scoping, community participation, analysis of alternatives, impact prediction, and impact evaluation. These deficiencies in the quality of EISs can be attributed to a range of factors as mentioned by the interviewees. The interviewees were asked the question "Why and when is the quality of EISs poor"? The factors influencing the quality of EISs in the context of Bangladesh (Figure 4.4) are analyzed here based on the information gathered from interviews.

4.8.1 Shortage of Study Time

Shortage of study time is one of the major factors for the poor quality of EISs, as determined by interviewees. Often proponents want to complete an EIA study as soon as possible. Sometimes proponents ask the consultants to complete the EIA study within 4 weeks. The findings of this study show that the average time for the study of EIAs is 4 months only, irrespective of the sectors. In some sectors, an EIA study requires at least 1 year in order to be able to view the seasonal variation and, thereby, assess the impacts on different environmental components. However, proponents do not want to wait this length of time (1 year).

4.8.2 Inadequate Baseline Data and Access to Data

One of the major problems of any EIA study in Bangladesh is that there is insufficient baseline information for social, physical, and ecological conditions, at both the national and local levels. The lack of publicly available and accurate data complicates the job of consultants. Even when and where data are available, they are shared on an ad hoc basis among several loosely connected agencies. Often the data are neither published nor cataloged. For example, public research institutions, such as the Soil Research Institute, the River Research Institute, and other specialized departments including non-governmental research organizations in Bangladesh, are reluctant to disseminate widely their survey results and findings. Furthermore, the public agencies are often reluctant to provide data or information to EIA consultants. The consequence of this is that the process of data collection becomes time consuming and costly.

In all of the categories of data, ecological data are found to be more deficient than any other category, such as social or physical data. This is due to a lack of ecological expertise and initiative among data collectors of different agencies. As in many other developing countries, the collection of ecological data is time consuming in Bangladesh. There is also the lack of available equipment for the collection of ecological data.

4.8.3 Attitude of Consultants and Proponents

The quality of EISs is often undermined by clandestine motives of consultants. The objective of an EIS is to care for the environment by uncovering the significant adverse effects of a development project. However, this objective is often undermined by the consultants. They often tend to serve the commercial interests of proponents. This can be substantiated by the findings of Momtaz (2002) in Bangladesh who stated that

> … their (the consultants) intention is to get an EIA done that would highlight the benefits and justify the proposal in order to obtain environmental clearance from the DOE or from the donor agencies for the purpose of fund clearance. It is therefore the job of the consultants to satisfy the proponent's requirements rather than carrying out the EIAs to ensure environmental and social sustainability of projects.
>
> *Momtaz (2002, p. 176)*

There is no code of ethics for the consultants in Bangladesh to control any unethical behavior, such as preparing EISs with false or misleading information. The opportunity for unethical practice by consultants is compounded when proponents do not have adequate experience or expertise in EIAs. Experience elsewhere, for example in China, shows that the existence of the provision of codes of conducts may improve this situation (Wang et al., 2003).

The proponents in Bangladesh often have the attitude that an EIA needs to be conducted to fulfill legal obligations or donor requirements and not necessarily to minimize and mitigate the potential impacts of the project. This myopic view is found to be acute particularly among the proponents in the private sectors where the proponents tend to look after their personal profits. Project proponents sometimes consider that getting an Environmental Clearance Certificate from the DOE is the end.

4.8.4 Lack of EIA Experts

Proponents in Bangladesh hire EIA consultants to prepare EIA reports. However, there is a shortage of EIA experts in Bangladesh. The reasons behind the lack of EIA experts in Bangladesh are multidimensional. The manpower trained in EIAs is not always available for the preparation of EISs. This is because they change their professional field and do not prepare EISs. Indeed, there is no scope to work as a full-time EIA expert, because the EIA, as a market, has not fully expanded compared to developed countries, such as Australia, the United States, and Canada. In the absence of available EIA experts, proponents often tend to prepare EIA reports by hiring junior or inexperienced EIA consultants. Proponents also hire junior consultants when they do not want to spend much money on an EIA. Due to a lack of expertise, consultants often fail to understand what information is required to conduct an EIA for a particular project.

4.8.5 Defective Service Procurement Process

The Service Procurement Act 2004 in Bangladesh states various methods for the procurement of consulting services in the public sector. Sometimes, the procurement methods prescribed by the Act are not followed properly by the public proponents. In this way, the defective service procurement processes may occur and, thus, compromise the quality of consultants and thereby the quality of EIA studies, particularly in the public sectors. In many cases, proponents invite Expressions of Interest (EOI) to conduct an EIA based on Quality and Cost-Based Selection or Least Cost Selection (LCS), ignoring the Fixed Budget (FB). Ideally the EOI should call for proposals from the consulting firms based on the FB method to study EIAs.

When the bidding is based on LCS, there is less opportunity to improve the technical quality of the study, since the lowest bidder tends to be awarded the contract. With this type of bidding process (LCS), there is a danger of the intrusion of inexperienced and opportunistic firms who have little to offer other than the lowest price to undertake the study.

4.8.6 Lack of Adequate Funds

The quality of an EIS also depends on the adequacy of funding allocated for the EIA study. However, "funding for EIS is still very limited even for big projects in Bangladesh. Still there is a lack of ideas among the proponents that EIA is a detailed work and needs adequate funds to perform tasks effectively and efficiently." Ideally 1–2.5% of the total project cost should be allocated for an EIA study (Modak and Biswas, 1999). In Bangladesh, the funds allocated for an EIA study is less than 1%.

4.8.7 Weak TOR

The majority of the informants said that the TOR prepared for the study of EIAs is often found to be very general and a prototype in nature in Bangladesh. This assumes no variation in TORs for EIA studies, irrespective of the sectors. The scope of the work to be performed by the consultants is described narrowly in this document. Often the TOR prepared for an IEE is found to be the same as for the full EIA. There are four aspects affecting the quality of TORs in Bangladesh: (1) proponents' inadequate capacity to prepare a good TOR; (2) inadequate time for conducting EIA allocated by the proponent; (3) lack of sector-specific EIA guidelines; and (4) proponents do not submit their TORs at the DOE for approval.

4.8.8 EIA Team

An EIA team plays a vital role in ensuring the quality of EIA reports. An EIA is an interdisciplinary study and, therefore, a team of EIA experts with various backgrounds covering the social, physical, and ecological areas is essential. However, in practice, proponents often only hire one individual to prepare an EIA report and this, therefore, affects the quality.

In Bangladesh, one of the reasons for the weak presentation and assessment of social impacts in EISs is the lack of involvement of social scientists. It has been observed that members of an EIA team are often dominated by consultants with backgrounds in physical sciences, including engineers and physical scientists. Ecologists or social scientist are nominally involved.

4.9 CONCLUSION

The purpose of this chapter was to investigate the quality of EISs in Bangladesh as one of the key aspects of an effective EIA system. The findings show that the quality of EISs in Bangladesh is only "just satisfactory." This means there are some major tasks that are not performed adequately. Also in this chapter, factors influencing EIS quality in the context of Bangladesh were identified and analyzed. Although the quality of EISs is a key indicator of an effective EIA system, even an EIS of excellent quality alone cannot represent a successful EIA regime. This requires adequate implementation of mitigation measures (as recommended by the EISs) at the post-EIS phase of EIA process. Chapter 5 examines to what extent the mitigation measures are implemented in order to minimize the environmental impacts of projects in Bangladesh.

APPENDICES

Appendix 4.1 List of EISs of Projects Reviewed

Sl. No.	Project ID	Project Name	Year
1	INDUSTRY-01/2000	EIA for CETP Installation on the Industrial Activities at Hazaribag Area, Dhaka	2000
2	INDUSTRY-02/2005	EIA of CETP and other industrial installations in the proposed Tannery Estate, Savar, Dhaka	2005
3	INFRASTRUCTURE-03/2007	Detail Design of Civil Works Under Road Network and Improvement project	2007
4	WATER-04/2006	EIA and SIA of Haor Rehabilitation Scheme (Haor Development and Rehabilitation Project) 2006	2006
5	WATER-05/2006	EIA and SIA for the improvement of navigation and environment of Korotoa River 2006	2006
6	WATER-06/2006	Hatya-Nijhum Dwip Cross Dam Project	2006

(Continued)

Appendix 4.1 List of EISs of Projects Reviewed—cont'd

Sl. No.	Project ID	Project Name	Year
7	WATER-07/2007	EIA and SIA of Flood Control Embankment and River Bank Protection on the Left Bank of Jamuna and Right Bank of Dhaleswari River	2007
8	ENERGY-8/2008	2D Seismic Acquisition Survey for the Project of Block 7	2008
9	ENERGY-9/2004	Construction of 25 km Gasp Pipeline From Maulivi Bazar to Rashidpur	2004
10	ENERGY-0/2000	EIA of Meghnaghat Power Plant Project-450 MW CC Gas Turbine Power Station Project	2000
11	ENERGY-1/2005	Nakla, Hatikumut-Bonara-Ishwardi-Veramara 101 km Gas Transmission Pipe Line Project	2005
12	ENERGY-2/2001	EIA of Bangabandhu Bridge Approach Road Gas Pipeline Project	2001
13	ENERGY13/2008	Bakharabad-Siddhirganj Gas Transmission Pipeline Project	2008
14	ENERGY-4/2008	Recompletion of Wells at Titas and Bakharabad Gas Fields and Installation of Gas Process Plant at Titas Gas Field	2008
15	ENERGY-15/1999	EIA for REJU-1 Offshore Exploration Drilling at Block 7	1999
16	INFRASTRUCTURE-16/1998	EIA of North Dhaka East Sewerage System 1998	1998
17	ENERGY-17/2008	EIA of Siddhirganj-Manikganagar 230 kV Transmission Line Project	2008
18	ENERGY-18/2001	Environmental Impact Assessment of gas project—Block 5	2001

Appendix 4.1 List of EISs of Projects Reviewed—cont'd

Sl. No.	Project ID	Project Name	Year
19	WATER–19/2007	EIA and SIA for Integrated Water Resource Management in Chalan Beel Area Development Project	2007
20	INDUSTRY–20/2001	EIA of American and Efrid Bangladesh Sewing Thread Dyeing and Finishing Plant Project	2001
21	WATER–21/1998	EIA of Nijhum Dwip Integrated Development Project	1998
22	INDUSTRY–22/2005	17 km Long Overland Elevated Belt Conveyer Project	2005
23	WATER–23/2005	Southwest Area Integrated Water Resources Planning and Management Project	2005
24	ENERGY–24/2008	EIA of 2×250 MW Gas Turbine Power Plant at Siddhirganj Project	
25	WATER–25/2001	Environmental Impact Assessment of Gorai River Restoration Project	2001
26	INDUSTRY–26/1997	Environmental Impact Assessment of Lafarge Surma Cement Project	2001
27	WATER–27/1998	EIA of of Khulna-Jessore Drainage Rehabilitation Project	1998
28	INFRASTRUCTURE–28/1990	Environmental Impact Assessment of Jamuna Multipurpose Bridge Project	1990
29	INFRASTRUCTURE–29/2007	Environmental Impact Assessment of Third Karnaphuli Bridge Project	2007
30	INFRASTRUCTURE–30/1996	EIA of Nalka-Hatikamrul-Bonpara New Road Project	1996

Appendix 4.2 A Sample of Collation Sheet

Name of Project: ....
Concerned Ministry/Organization: ....
Name of Consulting Firm: ....
Source of Finance (GOB/Donor Aided): ....
Year of Preparation: ....
Size of EIA Team: ....
Length of EIS: .... pages

Area	Grade Points	Area	Grade Points	Area	Grade Points	Area	Grade Points
1		2		3		4	
1.1		2.1		3.1		4.1	
1.1.1		2.1.1		3.1.1		4.1.1	
1.1.2		2.1.2		3.1.2		4.1.2	
1.1.3		2.1.3		3.1.3		4.1.3	
1.1.4		2.1.4					
1.1.5							
1.1.6							
1.1.7							
1.1.8							
1.2		2.2		3.2		4.2	
1.2.1		2.2.1		3.2.1		4.2.1	
1.2.2		2.2.2		3.2.2		4.2.2	
1.2.3		2.2.3		3.2.3		4.2.3	
1.2.4							
1.2.5							
1.3		2.3		3.3		4.3	
1.3.1		2.3.1		3.3.1		4.3.1	
1.3.2		2.3.2		3.3.2		4.3.2	
1.3.3		2.3.3		3.3.3			
		2.3.4		3.3.4			
1.4		2.4				4.4	
1.4.1		2.4.1				4.4.1	
1.4.2		2.4.2				4.4.2	
		2.4.3					
		2.4.4					
1.5		2.5					
1.5.1		2.5.1					
1.5.2		2.5.2					
1.5.3		2.5.3					
1.5.4		2.6					
1.5.5		2.6.1					
1.5.6		−2.6.4					

Appendix 4.3 Summary Results of the Quality of 30 EISs

Sl. No.	Project ID	Area#1	Area#2	Area#3	Area#4	Overall quality
1	INDUS-01/2000	D	F	F	E	E
2	INDUS-02/2005	B	D	C	B	C
3	COM-03/2007	C	B	B	B	B
4	WATER-04/2006	C	B	B	A	B
5	WATER-05/2006	B	B	A	A	A
6	WATER-06/2006	B	D	C	C	C
7	WATER-07/2007	C	C	C	B	C
8	ENERGY-08/2008	D	E	C	E	D
9	ENERGY-09/2004	A	C	D	B	B
10	ENERGY-10/2000	C	C	D	C	C
11	ENERGY-11/2005	B	C	B	B	B
12	ENERGY-12/2001	C	D	D	D	D
13	ENERGY-13/2008	D	E	E	D	D
14	ENERGY-14/2008	C	D	C	C	C
15	ENERGY-15/1999	C	E	C	C	D
16	URBAN-16/1998	B	D	C	B	C
17	ENERGY-17/2008	C	C	C	B	C
18	ENERGY-18/2001	E	F	E	D	E
19	WATER-19/2007	A	B	A	A	A
20	INDUS-20/2001	D	E	D	B	D
21	WATER-21/1998	C	C	C	D	C
22	INDUS-22/2005	C	D	C	D	D
23	WATER-23/2005	A	C	C	A	B
24	ENERGY-24/2008	A	B	B	A	A
25	WATER-25/2001	B	B	C	B	B
26	INDUS-26/1997	C	E	D	B	D
27	WATER-27/1998	C	C	B	B	C
28	COM-28/1990	B	D	C	C	C
29	COM-29/2007	C	D	B	B	C
30	COM-30/1996	B	B	B	A	B

Summary

Areas	Total "A"	Total "B"	Total "C"	Total "D"	Total "E"	Total "F"
Area#1	4	8	13	4	1	0
Area#2	0	7	8	8	5	1
Area#3	2	7	13	5	2	1
Area#4	6	12	5	5	2	0
Overall	3	6	11	8	2	0

REFERENCES

Ahammed AKMR, Harvey N. Evaluation of environmental impact assessment procedures and practice in Bangladesh. Impact Assess Project Appraisal 2004;22(1):63–78.

Asian Development Bank (ADB). Environmental assessment guidelines. Manila, The Philippines; 2003.

Badr E, Zahran AA, Cashmore M. Benchmarking performance: environmental impact statements in Egypt. Environ Impact Assess Rev 2011;31(3):279–85.

Barker A, Wood C. An evaluation of EIA system performance in eight EU countries. Environ Impact Assess Rev 1999;19:387–404.

Canelas L, Almansa P, Merchan M, Cifuentes P. Quality of environmental impact statements in Portugal and Spain. Environ Impact Assess Rev 2005;25:217–25.

Cashmore M, Christophilopoulos E, Cobb D. An evaluation of the quality of environmental impact statements in Thessaloniki, Greece. Journal of Environmental Assessment Policy and Management 2002;4(4):371–95.

Department of Environment (DOE). EIA Guidelines for Industries. Ministry of Environment and Forest, Government of Bangladesh, Dhaka; 1997.

Department of Environment (DOE). Personal Communication; 2009, dated 25/7/2009.

Formby J. The politics of environmental impact assessment. Impact Assess Bull 1991;8:1–2.

Glasson J, Therivel R, Weston J, Wilson E, Frost R. EIA-Learning from experience: changes in the Quality of Environmental Impact Statements for UK Planning Projects. J Environ Plan Manage 1997;40(4):451–64.

Government of Bangladesh (GOB). Khulna-Jessore Drainage Rehabilitation Project, Bangladesh Water Development Board. Ministry of Water Resource Development, Government of Bangladesh; 1998.

Government of Bangladesh (GOB). Meghnaghat Power Plant Project. Ministry of Energy, Power and Mineral Resources, Government of Bangladesh; 2000.

Government of Bangladesh (GOB). EIA of Bangabandhu Bridge Approach Road Gas Pipeline Project. Ministry of Energy, Power and Mineral Resources, Government of Bangladesh; 2001.

Government of Bangladesh (GOB). EIA and SIA of Haor Rehabilitation Scheme. Dhaka, Bangladesh: Ministry of Water Resources; 2006.

Government of Bangladesh (GOB). Third Karnaphuli Bridge Project. Road and Highway Division, Ministry of Communication, Government of Bangladesh; 2007.

Gray I, Edward-Jones G. A review of environmental statements in the British forest sector. Impact Assess Project Appraisal 2003;21(4):303–12.

Hickie D, Wade M. Development of guidelines for improving the effectiveness of environmental assessment. Environ Impact Assess Rev 1998;18:267–87.

Hughes J, Wood C. Formal and informal environmental assessment reports - their role in UK planning decisions. Land Use Policy 1996;13(2):101–13.

Jamuna Multipurpose Bridge Authority (JMBA). Resettlement in JMBP: Assessing Process and Outcomes. Qualitative evaluation of RRAP (Revised Resettlement Action Plan) and Project on EFAP (Erosion and Flood Affected Persons), Government of Bangladesh, Dhaka; 2001.

Kabir SMZ, Momtaz S. The quality of environmental impact statements and environmental impact assessment practice in Bangladesh. Impact Assess Project Appraisal 2012;30(2):94–9.

Lee N, Colley R. Reviewing the quality of environmental assessments. Occasional paper, Number-24, EIA CentreUK: University of Manchester; 1992.

Lee N, Colley R, Bonde J, Simpson J. Reviewing the quality of environmental assessments and environmental appraisals. Occasional paper 55, EIA Centre UK: University of Manchester; 1999.

Lockie S, Momtaz S, Taylor B. Meaning and construction of social impacts: water infrastructure development in Australia's Gladstone/Calliope region. Rural Soc. 1999;9(3):529–42.

Modak P, Biswas AK. Conducting environmental impact assessment for developing countries. New York: United Nations University Press; 1999.

Momtaz S. Environmental impact assessment in Bangladesh: a critical review. Environ Impact Assess Rev 2002;22:163–79.

Momtaz S. The practice of social impact assessment in a developing country: the case of environmental and social impact assessment of Khulna-Jessore Drainage Rehabilitation Project in Bangladesh. Impact Assess Project Appraisal 2003;21(2):125–32.

Momtaz S. Institutionalizing social impact assessment in Bangladesh resource management: limitations and opportunities. Environ Impact Assess Rev 2005;25:33–45.

Pinho P, Maia R, Monterroso A. The quality of Portuguese environmental impact studies: the case of small hydropower projects. Environ Impact Assess Rev 2007;27:189–205.

Sandham LA, Pretorius HM. A review of EIA report quality in the North West province of South Africa. Environ Impact Assess Rev 2008;28(4–5):229–40.

Steinmann A. Rethinking health impact assessment. Environ Impact Assess Rev 2008; 20:627–45.

Stolp A, Groen W, Vliet JV, Vanclay F. Citizen value assessment: incorporating citizen's value judgement in environmental impact assessment. Impact Assess Project Appraisal 2002;20(1):11–23.

Sullivan WC, Kuo FE, Prabhu M. Assisting the impact of environmental impact statements on citizens. Environmental Impact Assessment Review 1996;16:171–82.

Wang Y, Morgan RK, Cashmore M. Environmental impact assessment of projects in the People's Republic of China: new law, old problem. Environ Impact Assess Rev 2003;23(5):543–79.

Wende W. Evaluation of the effectiveness and quality of environmental impact assessment in the Federal Republic of Germany. Impact Assess Project Appraisal 2002;20(2):93–9.

Weston J. Reviewing environmental statements: new demands for the UK's EIA procedure. Plan Pract Res 2000;15(1/2):135–42.

World Bank. The World Bank operational manual (OP 4.01). Washington, DC, USA; 1999.

CHAPTER 5

Evaluating Social Impact Assessment

Contents

5.1 INTRODUCTION

Developing countries were prompt in embracing environmental impact assessment (EIA) in the planning process after its introduction in the United States through the National Environmental Policy Act of 1969 (NEPA 1969) and its institutionalization in developed countries. While EIA is regarded as the forerunner of social impact assessment (SIA), the latter remained a poor cousin of EIA until recently. SIA is the process of assessing and managing the consequences of development projects, policies, and decisions on people. Both have their origin in NEPA 1969. The main objective of the Act was to identify the intended and unintended effects of planned interventions on communities in order to ensure project sustainability (Burge, 2003). Despite the same origin, SIA emerged as a separate field of applied social science due to the realization in the early days of NEPA that

EIAs could not adequately address social issues. In the late 1970s, many developed countries and some developing countries adopted SIA as a means of addressing social issues arising from development initiatives. However, SIA remained basically an integral component of EIA and is yet to be as firmly established in environmental planning as EIA, especially in developing countries (Cox et al., 2000). Due to its importance in a country like Bangladesh where densely populated villages and localities are often facing the effects of development activities, this chapter examines the emergence of SIA within the framework of EIA in Bangladesh. It evaluates the place of SIA in project planning and implementation by major agencies.

5.2 CONCEPTUALIZING SIA: ORIGIN, EVOLUTION, AND DEFINITION

The issue of social impacts was first identified by an Inuit elder when he referred to the proposed Trans-Alaska pipeline project, the work that went into identifying its impacts on the permafrost and the caribou population and said "Now that we have dealt with the problem of the permafrost and the caribou and what to do with hot oil, what about changes in the customs and ways of my people?" (Vanclay and Bronstein, 1995). The project was a huge undertaking in the construction of 1288-km (800 miles) long pipeline to carry oil from the North Slope of Alaska to the northern most ice-free port—Valdez, Alaska (Trans-Alaska Pipeline project Web site). It would involve almost 42,000 workers in the project. As Alaska's population was small, most of the workers would come from the mainland United States. This influx of workers with different dialect and culture would have significant impact on the cultures of the local Inuit people. The term "social impact assessment" is believed to have been used for the first time in 1973 in the EIA documents of Trans-Alaska Pipeline project to refer to impacts on the culture of Inuit people.

The first official document on SIA guidelines titled "Guidelines and Principles for Social Impact Assessment" was prepared in the United States by the Interorganizational Committee on Guidelines and Principles for Social Impact. Although NEPA 1969 came into force in January 1970, it was not until 1994 that the United States had its first SIA guidelines. The agencies, planners, and decision-makers who were involved in the implementation of NEPA 1969 recognized the need for better understanding of social issues related to development projects. This recognition led to the formation of the Committee which submitted the report in 1994 (Interorganizational Committee on Guidelines and Principles, 1994).

The Committee defined SIA as

to assess or estimate, in advance, the social consequences that are likely to follow from specific policy actions (including programs and the adoption of new policies), and specific government actions (including build buildings, large projects and leasing large tracts of land for resource extraction), particularly in the context of the U.S. National Environmental Policy Act of 1969.

Interorganizational Committee on Guidelines and Principles (1994, p. 1)

In a nutshell, the idea is to identify the positive and negative impacts so that positive impacts can be enhanced, long-term irreversible negative impacts can be avoided, acceptable negative impacts can be minimized, and appropriate mitigation and compensation measures can be developed for a project to be socially sound.

International Association for Impact Assessment (IAIA, 2003) published its SIA guidelines in 2003. According to the document, SIA includes

the processes of analysing, monitoring and managing the intended and unintended social consequences, both positive and negative, of planned interventions (policies, programs, plans, projects) and any social change processes invoked by those interventions. Its primary purpose is to bring about a more sustainable and equitable biophysical and human environment.

IAIA (2003, p. 1)

5.2.1 Recent Conceptual and Methodological Advances in SIA

As mentioned above, SIA has come a long way since its emergence in the 1970s as a social planning tool. The international development partners, such as World Bank Group, Asian Development Bank (ADB), and United Nation Development Program, have embraced SIA and implemented it in their projects and programs in developing countries. There have been numerous publications on conceptual and methodological aspects of SIA in journals like "Impact Assessment and Project Appraisal" and "Environmental Impact Assessment Review." Developed and developing countries have further strengthened SIA by way of legal provisions. However, it is widely believed that the benefits of SIA have not been fully achieved. This is partly due to the fact that SIA has been implemented as a textbook exercise to identify negative impacts of development projects to formulate mitigation and management plans in order to meet regulatory requirements and partly because of the fact that "the SIA community has failed to convince all its stakeholders of the full potential of SIA" (Vanclay and Esteves, 2011, p. 3). SIA can be much more than a step-by-step process for identification of impacts and fulfillment of legal requirements. It can be a mechanism for promoting social sustainability, for example, through facilitating negotiation processes between mining companies and local indigenous populations; assisting local communities in understanding the impacts and benefits of a project; helping companies and communities to manage social change processes for maximizing project benefits and minimizing their negative effects. In the end, SIA can be a tool for positive social change (Vanclay and Esteves, 2011). Furthermore, the recent developments in the practice of SIA and advancement in the conceptual framework suggest that SIA would play a wider role by considering gender inequality (Lahiri-Dutt and Ahmad, 2011), ethical issues (Baines and Taylor, 2011), human rights (MacNaughton and Hunt, 2011), indigenous people (O'Faircheallaigh, 2011), cumulative social impacts (Franks et al., 2011), climate change and disaster (Cottrell and King, 2011), and local procurement (Esteves et al., 2011).

5.3 INSTITUTIONALIZING SIA IN BANGLADESH

EIA is a relatively newcomer in the development arena of Bangladesh. Bangladesh National Environmental Policy 1992 (GOB, 1992) sets out the basic framework for environmental action in Bangladesh and paves the way for EIA for all major project interventions. As stated earlier (see Chapter 3), the country had its first EIA guidelines in 1992 for infrastructure development in water sector. Despite weaknesses in the implementation side, EIA has become a well-recognized part of policy development (Momtaz, 2002).

5.3.1 Legal Framework of SIA

A number of legal documents played important role in the shaping of consideration of the social issues in development activities. National Environmental Policy 1992 (GOB, 1992) is one of the early policies related to environment in Bangladesh. The first objective of the NEP is to ensure sustainable development through balanced environmental protection and overall development. It shows that the NEP is not just about the biophysical system but it is also about a balanced development for the benefit of the society. National Environmental Management Action Plan 1995 was developed by the Ministry of Environment and Forest (MoEF, 1995) through a consultative process. This is the government's action plan for all major sectors of the country. It identifies various issues for all sectors of the economy, highlights people's concerns for those issues, and proposes management recommendations for those issues. The Environment Conservation Act 1995 was promulgated to provide for conservation of the environment, improvement of environmental standards, and control and mitigation of environmental pollution.

The laws Environmental Conservation Act 1995 (ECA '95) and Environmental Conservation Rules 1997 (ECR '97) govern the conduct of EIA in Bangladesh (see Chapter 3). Department of Environment (DOE) is the lead government agency that regulates project interventions and reviews environmental impact statements for the purpose of issuing environmental clearance certificates to proponents (Momtaz, 2002). There is no mention of SIA in any of the legal documents (BCAS, 1999; DOE, 1997). Rather, the DOE's definition of the term "environment" has been expanded to include human issues "the inter-relationship existing between physical properties of earth (water, air, soil) and living organisms (human beings, plants, micro-organisms)" (BCAS, 1999, p. 39). DOE's *EIA Guidelines for Industries* is also inclusive as far as social issues are concerned. It explains "A comprehensive EIA . . . involves study of the probable changes in the physical and biological as well as socio-economic environment which may result from the proposed development activity or project . . ." (DOE, 1997, p. 2). One of the three major components in the checklist of environmental components is "Human" component which includes "Health and Safety," "Social and Economic," and "Aesthetic and Cultural" aspects (DOE, 1997, p. 68). Special emphasis on social issues can be found again in the section outlining various steps in EIA where public participation has been identified as an important ongoing activity in the assessment procedure: "Since the general public is the ultimate recipient of the economic benefits and environmental damages, an EIA study should involve the public as part of the decision-making process" (DOE, 1997, p. 34). The DOE (1997) further suggests that as many people as possible should be contacted to ensure effective participation and all possible means of communication, e.g., radio, television, news releases, newsletters, sample polls, lobbying, workshops, public meetings, public hearings, information van, and citizens' advisory committee, should be adopted to achieve maximum participation. This anthropocentric view of EIA indicates that assessments are carried out in order to minimize the effects of "environmental damage" on humans thereby placing significant importance on human factors of planned intervention. Table 5.1 lists the major events in the evolution of social issues in EIA in Bangladesh (Momtaz, 2006).

According to the DOE's EIA Guidelines for Industries, "social and economic" components of the environment include employment, housing,

Table 5.1 Major events in the evolution of SIA in Bangladesh

Organization	Date	Event/document	Status of social assessment
Flood Action Plan (FAP)	1992	First EIA guidelines	Incorporation of peoples' participation in entire project cycle
Local Government Engineering Department (LGED)	1992	EIA guidelines for small scale water resource development project	Identification of social environment and human interest as major impact area
Government of Bangladesh (GOB)	1995	First EIA legislation ECA, 1995	Consideration of socioeconomic impacts
Department of Environment, GOB	1997	Environment conservation rules to enforce ECA, 1995	Consideration of human settlement as ecologically critical area
EGIS, Ministry of Water Resources, GOB	1998, 2001	Major EIA/SIAs	Participatory rural appraisal used as the basis for SIA
CARE Bangladesh and USAID	1998	EIA Field Handbook for Rural Road Improvement Projects	Impact on human interest has been identified as a significant area of investigation in EIA
Asian Development Bank	1999	Environment operational strategy	Initial social impact assessment of development project
Water Resource Planning Organization (WARPO)	2001	Major review of FAP (1992) and EIA guidelines	Making people's participation (PP) mandatory. Recognition of PP as the key to achieving long-term success and sustainability of projects

Source: Compiled by the authors from relevant original documents.

education, utilities, and amenities; and "aesthetic and cultural" components include landform, climate, biota, man-made objects, historic or religious places and structures and wilderness, water quality, atmospheric quality, tranquility, sense of community, community structure, and landscape. These are comparable with the broad SIA variables as suggested by the

Interorganizational Committee on Guidelines and Principles (1994, p. 15): population characteristics, community and institutional structures, political and social resources, individual and family changes, and community resources.

5.3.2 Organizational Framework of SIA

As discussed before, SIAs are usually conducted as part of EIA, and as such the organizational framework of SIA is no different to that of EIA as discussed in Chapter 3.

International development partners and local and international nongovernmental organizations (NGOs) are playing important roles in the development of EIA, SIA, and community participation guidelines and implementation of EIA. In collaboration with responsible government agencies, they contributed to the refinement of procedures and acted as a safeguard in quality assurance in the EIA sector especially in the projects funded by them. Some examples of their activities in relation to EIA and SIA have been listed in Table 5.2.

(a) *Department of Environment*: As stated earlier, ECA 1995 provides the legal authority to the DOE to implement and enforce EIA rules and regulations. It has the authority to review EISs.

(b) *DOE Steps in SIA*: There is no separate procedure to follow in the conduct of SIA; however, most EISs would have a socioeconomic impacts chapter to identify social and economic impacts of a proposed development. The important stages in the development of an EIA study, according to EIA Guidelines for Industries prepared by DOE (1997), have been discussed in Chapter 3.

5.3.3 Status of SIA in NGOs and Development Partners

(a) *SIA in the Center for Environmental Geographic Information Systems*: The Government of the People's Republic of Bangladesh established the Center for Environmental and Geographic Information Services (CEGIS) as a public trust in 2002 under the Trusts Act of 1982. It has been functioning under the aegis of the Ministry of Water Resources and a Board of Trustees. This organization issued the country's first EIA guidelines in 1992 for physical interventions in water sector (FAP, 1992). While the main focus of EIA was on biophysical environment, the four areas that were considered also

Table 5.2 National and international organizations and their EIA/SIA-related activities

Organization	Activities
LGED—government agency	- Rural infrastructure development in collaboration with CARE and USAID - Developed EIA/SIA guidelines - Published "Guidelines on Environmental Issues Related to Physical Planning" in 1992 with emphasis on humans - EIA guidelines (LGED, 2007)
WARPO—government agency	- Conducted a major review of the first EIA guidelines on behalf of the government - Made public participation mandatory
CEGIS—government agency	- Issued the country's first EIA guidelines in 1992 for physical interventions in water sector (FAP, 1992) - Conducted major EIA/SIA studies (CEGIS, 2008)
CARE Bangladesh—international NGO	- Introduced "Food for Work" program for involvement of rural poor in rural road construction - Conducted EIA of rural road projects - Developed EIA guidelines for these studies
The World Bank Group—development partner	- Involved in EIA in Bangladesh since its inception - All WB-funded projects under categories A and B (environmentally significant) are subject to environmental and social analysis and impact assessment - Provides EIA training
Asian Development Bank—development partner	- Its role is similar to that of WB - Developed guidelines for Bangladesh EIA and SIA (ADB, 2007) - Had major involvement in the construction of "Jamuna Multipurpose Bridge"—country's largest infrastructure project - Provides EIA training

include socioeconomic issues in addition to land and water resources. FAP (1992) guidelines were reviewed in 2001 (WARPO, 2011). The guidelines do not recognize SIA as a separate process. But the reviewed document indicates "The guiding principle behind the Guidelines is to safeguard the physical, biological and socio-economic environments during project preparation and operation" (WARPO, 2011, p. 1). A revised EIA procedure is proposed in the reviewed guidelines with more emphasis on people's participation which is regarded as the principal means "to ensure that the groups affected by a proposed project have the opportunity to decide whether or not it should be implemented and—if it is to go ahead—to modify it if necessary" (WARPO, 2011, p. 13) and to "ensure early detection of possible social conflicts arising from the proposed interventions, enhancements and mitigations and to explore ways of minimising them—eg through negotiation and education" (WARPO, 2011, p. 14). Chapter 6 further investigates into community participation in EIA and SIA.

EGIS has conducted a number of major SIAs of river restoration and drainage rehabilitation projects in the past few years. The latest being "Environmental and Social Impact Assessment of Flood Control Embankment 2006." These SIAs were conducted as part of EIAs of those projects. This organization has developed a stakeholders' consultation manual that testifies to the prominence of social issues in environmental planning. CEGIS adopted World Bank and ADB EIA/SIA procedures as both these organizations funded the studies. A review by Momtaz (2003) of the EIA/SIA reports for these projects reveals that they satisfy most of the review criteria developed by Modak and Biswas (1999). It will be found in the case study chapter that despite some good guidelines and reports, there are major issues with implementation of these guidelines and the implementations of the report recommendations.

(b) *World Bank and Its Environmental Strategy*: Since 1989, the World Bank staff are required to screen all proposed new investment projects with respect to their potential environmental impacts and to classify them accordingly. In a sense that was the beginning of EIA in WB projects in recipient countries. A Social Development Task Group was formed in 1996 to consider how social issues in the bank-funded projects could be integrated in project cycle. Later, a Social Development Board was established. Today, all WB-funded projects under categories A and B (environmentally significant) are subject to environmental and social analysis and impact assessment.

(c) *Asian Development Bank*: Bangladesh is the 20th largest shareholder in ADB as well as one of the major recipients of ADB development loans. According to ADB, the principal goal of its activities in developing countries is to reduce poverty in developing member countries. ADB had a major involvement in the construction of Jamuna River Bridge—the largest infrastructure project in the history of Bangladesh (this is one of the case studies investigated for this book, see Chapter 7).

Recently, ADB has published "ADB Handbook on Social Analysis" (ADB, 2007). According to ADB, "the purpose of social analysis during project preparation is to identify opportunities, constraints, and risks for poor and marginalized groups related to the project; to establish a participatory process for the design of the project; and to prepare design measures to achieve social development outcomes and avoid or mitigate any social risks during implementation." ADB's concept of social analysis goes beyond the conventional notion of SIA that identifies positive and negative impacts of projects on affected people; it recognizes the fact that impacts are not ubiquitous on all affected groups and marginalized people are more vulnerable to project impact than any other groups. ADB, therefore, focuses specifically on poor people so that their rights are protected.

5.4 PERFORMANCE OF THE BANGLADESH SIA SYSTEM

Throughout the evolution phase of EIA, authors have attempted to determine if EIA was making a real impact on environmental protection and whether it was fulfilling the desired objectives for which it was introduced in the first place. Authors have used various criteria to evaluate EIA system of a country and also made comparative analysis of EIA performance in Europe, Africa, and Asia (Canelas et al., 2005; Cashmore et al., 2002; Glasson et al., 2005; Sandham and Pretorius, 2008). There have been very few studies on the performance of SIA system of a country (Ahmadvand et al., 2009; Androulidakis and Krakassis, 2006). Bangladesh has seen a number of studies reviewing the performance of EIA (e.g., Ahammed and Harvey, 2004; Momtaz, 2002). However, no systematic review has yet been done to assess the performance of SIA. In the following paragraphs, we have made an attempt to systematically evaluate the performance of Bangladesh SIA system. Five sets of criteria have been developed to assess the SIA system of Bangladesh (Tables 5.3–5.7). These criteria have been developed from a review of all aspects of EIA system in Bangladesh (see Chapter 3) and the literature that evaluated EIA and SIA systems in different developing countries (Ahmadvand et al., 2009; Androulidakis and Krakassis, 2006). The criteria enable us to assess the most important aspects of an SIA system, i.e., the adequacy of SIA legislation, administrative arrangements, practice, foundational measures, and the quality of SIAs. Tables 5.3–5.6 provide findings on the first four aspects and Table 5.7 focuses on the quality of SIAs—the last criterion.

Table 5.3 Assessment of the adequacy of SIA legislation in Bangladesh

Evaluation criteria	Assessment of the Bangladesh SIA system
Legal provision for SIA	ECA '95 provides the legislative basis with its expanded definition of the environment. No specific legislation on SIA
Provision for appeal by the developer or the public against decisions	Clause 8 and 14 of ECA '95 have clarified the right to appeal
Legal or procedural specification of time limits	ECA '95 provides clear instructions about maximum time period for various decisions which vary from 30 to 60 days
Formal provision for SEA	The concept of SEA is still in its early stage in Bangladesh as suggested in discussions with officials

Table 5.4 Assessment of the adequacy of the administrative arrangements for the conduct of SIA in Bangladesh

Evaluation criteria	Assessment of the Bangladesh SIA system
Existence of a competent authority for EIA/SIA and procedures for the determination of social acceptability	DOE is competent as the approval authority, but there are issues with resource allocation for DOE. DOE officials have acknowledged this
Existence of a review body for EIA/SIA	DOE is the statutory review body
Specification of the responsibilities of sectoral authorities in the EIA process	Responsibilities are devolved under different directorates within DOE, but issue of resource allocation remains
Level of coordination with other planning and management agencies	There seems to be lack of coordination between DOE and other organizations involved in EIA/SIA such as LGED, CEGIS, and the international development partners. This has been indicated by officials in CEGIS, CARE, ADB, and WB

5.4.1 Examples of Some Recent SIA of Major Projects in Bangladesh

The following projects have been chosen from the projects we have investigated for the purpose of reviewing quality of EISs.

(a) *Phulbari Coal Project (Asia Energy Corporation, 2006)*: This EIA is for a proposal for an open-cast mine to cover an area of 2180 ha with mining foot print of 5192 ha. It is expected that the project will provide more than

Table 5.5 Assessment of the adequacy of SIA practice in Bangladesh

Evaluation criteria	Assessment of the Bangladesh SIA system
Specified screening categories	DOE follows normative screening process—according to their potential environmental impacts, development activities have been classed under four categories (Green, Amber A, Amber B, and Red). ECR'97 provides a list of activities under each category
Systematic screening approach	Existing list of activities under various categories helps in screening process. Donor guidelines complement this process
Systematic scoping approach	DOE develops a list of terms of reference for the proponent once a development proposal is submitted. The terms of reference are then used as a guideline in scoping. However, often there are issues as to what should and should not be included in the scoping processes (Momtaz, 2002)
Requirement to consider alternatives	DOE guidelines do not say much about consideration of alternatives (DOE, 1997). However, an examination of EIAs of some major projects reveals that alternatives have been considered including no-development option. This is due to the influence of donor agency guidelines that emphasize this important aspect (e.g., ADB, 2007)
Specified content of SIA report	There are no specified contents of SIA as it is done as part of EIA. There is specified list of contents of EIA which includes description of social, environment, and socioeconomic impacts (DOE, 1997)
Public participation in SIA process	Community participation is a requirement in guidelines (DOE, 1997). Significant emphasis in donor documents (WB, 2002)
Requirement for social and environmental management plans	EMP is a requirement. Management of social issues is usually included in EMP as revealed during EIS analysis
Requirement for mitigation of impacts	Mitigation measures are a requirement and part of all EIA reports. However, there are issues with actual implementation. This information is obtained from interviews of practitioners. No systematic study on this yet

Table 5.5 Assessment of the adequacy of SIA practice in Bangladesh—cont'd

Evaluation criteria	Assessment of the Bangladesh SIA system
Requirement for impact monitoring	Monitoring measures are a requirement and part of all EIA reports. However, there are issues with actual implementation. In the absence of EIA follow-up study, it is difficult to determine the quality of monitoring. This information is obtained from interviews of practitioners
Experience of strategic environmental assessment	SEA is a new concept in Bangladesh as is cumulative impact assessment

Table 5.6 Assessment of the adequacy of foundational measures in the Bangladesh SIA system

Evaluation criteria	Assessment of the Bangladesh SIA system
Existence of guidelines	There are now guidelines for EIA which also includes social impact guidelines for major sectors; donor agencies have played an important role in the development of these guidelines (ADB, 2007; DOE, 1997; LGED, 2007)
SIA/EIA system implementation monitoring	There is no formal process in place to evaluate EIA/SIA implementation
Expertise in conducting SIA	Bangladesh has developed a good skill base in EIA and SIA. However, there is no mechanism in place for quality control in private sector development
Training and capacity building	Donor agencies offer training on a regular basis and have made significant contribution in capacity building

20,000 jobs directly and indirectly and contribute about 1% to the country's GDP. The proposal will lead to the displacement of population with at least 160 households needing relocation. The EIA study developed various social plans in response to indentified social impacts. They are resettlement and compensation plan, livelihood restoration plan, an indigenous people's development plan, conceptual land use plan for the extended township and surrounding villages, and public consultation and disclosure plan. The report

Table 5.7 Quality of social impact statements in Bangladesh

Attribute	Project 1	Project 2	Project 3	Project 4	Project 5	Project 6	Project 7	Project 8	Project 9	Project 10	Project 11
Executive summary	Good	Good	Poor	Poor	Good	Good	Good	Good	Absent	Good	Good
Description of the environment											
Description of natural environment	Good	Good	Good	Good	Good	Good	Good	Good	Good	Good	Good
Description of social environment	Good	Good	Good	Poor	Good	Good	Good	Poor	Good	Good	Good
Reference to existing similar project in the area											
Description of the project											
Description of proposed project	Good	Good	Good	Poor	Good	Good	Good	Good	Good	Good	Good
Description of operation of the project	Good	Good	Medium	Poor	Good	Good	Good	Good	Good	Good	Good
Justification of project' usefulness	Good	Good	Good	Medium	Good	Good	Good	Good	Good	Good	Good

Continued

Impact identification and prediction methods

Adequacy of impact identification method	Good	Good	Good	Medium	Good	Good	Good	Good	Good	Good	Good
Estimation of positive and negative social impacts	Good	Good	Medium	Poor	Good	Good	Good	Good	Good	Good	Good
Estimation of residual social impacts	Good	Good	Poor	Poor	Medium	Poor	Medium	Poor	Poor	Medium	Medium
Clarity and precision of the prediction	Good	Medium	Good	Poor	Medium	Good	Good	Good	Good	Good	Good
Mitigation and monitoring											
Discussion of proposed mitigation	Good	Good	Medium	Poor	Good	Good	Good	Good	Good	Good	Good
Appropriateness of mitigation	Good	Good	Medium	Poor	Good	Good	Good	Good	Good	Good	Good
Monitoring arrangements	Good	Good	Medium	Medium	Good	Good	Good	Good	Good	Good	Good
Environmental management plan	Good	Good	Medium	Medium	Good	Good	Good	Good	Good	Good	Good

Table 5.7 Quality of social impact statements in Bangladesh—cont'd

Attribute	Project 1	Project 2	Project 3	Project 4	Project 5	Project 6	Project 7	Project 8	Project 9	Project 10	Project 11
Alternatives											
Reference to alternatives	Good	Good	Poor	Poor	Good	Good	Medium	Good	Poor	Good	Good
Reference to procedures for selection among alternatives	Good	Good	Poor	Poor	Good	Good	Medium	Good	Poor	Good	Good
Comparison to a "no-action option"	Medium	Good	Poor	Poor	Good	Good	Absent	Good	Absent	Good	Medium
Public participation											
Contribution of public to the development of SIA	Good	Good	Medium	Poor	Good	Poor	Medium	Good	Good	Good	Good
Appropriateness of consultation methods	Good	Good	Medium	Poor	Good	Poor	Medium	Good	Good	Good	Good

indicates a thorough stakeholders' consultation process. However, it is not clear if the consultation outcomes were incorporated in the final decision making.

(b) *Siddhirganj-Maniknagar 230 kV Transmission Lines Project (CEGIS, 2008)*: This is a WB-funded project, 11-km long power lines. It will affect 82 households and some local organizations. As part of the EIA, an SIA was conducted and a resettlement action plan was developed. The SIA/RAP report spent several pages to identify how many households are going to be affected by its 42 towers. Only three pages are dedicated to analyzing social impacts, the most important part of an SIA. The SIA was conducted by an organization with reputation in SIA. However, there was no sociologist or anthropologist included in the study team. The report indicates that public consultation was conducted through focus group discussions. The study team also conducted individual interviews with affected industries and real estate companies. The report lists the mitigation measures suggested by the participants and indicates that the suggestions were taken into consideration during the development of the management plan.

(c) *Road Network Improvement and Maintenance Project: EIA (Louis Berger Group, 2007)*: This EIA was prepared by private consultants from India and Bangladesh for the Government of Bangladesh for a 204-km road network upgrade project. The upgrade project was funded by the ADB. It fell under the category Red and category A by DOE and ADB categorizations, respectively, for the purpose of EIA. Therefore, it required full EIA and was conducted in compliance with the GOB and ADB environmental guidelines, rules, and regulations. The study identified the social impacts using questionnaire surveys and field observation. Community consultation, focus group study, and meetings with groups were also employed as a means of collecting data about the potential socioeconomic impacts of the project. The surveys and discussions focused on various social aspects including impacts on the economic development, agricultural production, social and cultural resources, health, education, and sites of cultural significance.

(d) *EIA of Bridge Improvement Project (SAPROF, 2007)*: The report represents the EIA for seven bridges of the proposed Bridge Improvement Project (BIP). It was done on behalf of the Government of Bangladesh for the funding body Japan Bank for International Cooperation. The report used environmental scoping matrix to identify impact on the social environment and determine their magnitude using weighting (i.e., from $+3 =$ significant positive impact to $-3 =$ significant negative impacts). The report does have a brief "public consultation chapter" and claim to have discussed the issues in some roadside meetings. Detailed descriptions of social impacts and their mitigation measures are not provided. Nor did the report provide detailed descriptions of public meetings. This is one of the poor EIAs (with very thin socioeconomic section) reviewed by us. It was conducted by a local consultant team. The report did clearly outline the EIA requirements of the Japan Bank. However, it must have required approval from DOE and therefore should have fulfilled the minimum requirements.

5.4.2 Quality of SIAs

Qualities of EISs have been thoroughly evaluated in various countries and a set of rigorous procedures have emerged in recent literature (Canelas et al., 2005; Cashmore et al., 2002; Glasson et al., 2005; Lee and Colley, 1992; Sandham and Pretorius, 2008). We have also reviewed the quality of EISs in Bangladesh (Chapter 4). SIAs have not received similar scrutiny due mostly to the fact that often social impacts are incorporated as a part of EISs and the same criteria used for EISs apply to the sections on socioeconomic impacts. However, some authors (Ahmadvand et al., 2009; Androulidakis and Krakassis, 2006) have attempted to evaluate the status of SIA in developed and developing countries and made efforts to systematically examine the quality of relevant chapters on social impacts of EISs.

The DOE (1997) has developed a long list of issues to be examined in the review process of EIAs for environmental clearance to proposed projects. The list applies to the whole of an EIS including its socioeconomic chapter which is often regarded as social impact statement of the document. A number of issues directly relate to social impacts of a project, namely, "relevant social and policy issues of local nature relating to the project" and "significant social impacts related to the effects on particular interest groups." Table 5.7 examines the quality of SIA part of some major recent EISs in Bangladesh using evaluation criteria based on DOE review list and recent literature (Ahmadvand et al., 2009; Androulidakis and Krakassis, 2006). We have examined 11 EISs (Appendix 5.1) of major development projects for the past 10 years to determine if social impacts have received due considerations. These are among the EISs used for review in Chapter 4. It may be noted here that the conduct of separate SIAs (that constitute a part in the EISs) is a recent phenomenon in Bangladesh and the 11 EIAs reviewed here for their quality represent almost all major SIAs conducted in Bangladesh. DOE requirements and guidelines for a standard EIS were kept in mind while developing criteria for this assessment (DOE, 1997 and DOE, 2009 amendment of ECA 97). For each criterion, a three-point qualitative rating scale of "good," "medium," and "poor" was used. The term "absent" has been used if a criterion is totally missing. Scoring was based on a coding protocol adapted from recent literature (Ahmadvand et al., 2009; Androulidakis and Krakassis, 2006) and modified to suit Bangladesh context (Appendix 5.2).

Table 5.7 shows that the quality of SIAs is generally good. Out of 11 SIAs, only 2 have rating of "medium" and "poor" for most criteria. Others have rating of "good" for almost all aspects of SIA quality. Some 19 criteria have been used to rate each SIA. For 11 SIAs altogether, there are 209 ratings of "good," "medium," "poor," and "absent." Out of 209, 154 (74%) are "good," 25 (12%) are medium, and 27 (13%) are poor (Table 5.7).

5.5 DISCUSSION AND RECOMMENDATIONS

5.5.1 Adequacy of Legislative Arrangement of SIA

Since the independence in 1971, Bangladesh has been experiencing major development activities. Many of these activities are funded by international agencies and NGOs with poverty alleviation objectives. As a result, social concerns are often central to EIAs conducted for these projects. Social assessment has been an integral part of EIA since the early days of its inception in Bangladesh. FAP guidelines (LGED, 1992) included social aspects as one of the four major environmental components. Although SIA does not have separate legal status, it is well placed in project proposals.

Environmental Conservation Act and Rules, DOE guidelines, as well as the guidelines of the donor agencies played important role in the incorporation of human factor in EIA in Bangladesh. However, since SIA requirement is not separately mentioned in ECR '95 and ECA '97, there is a danger— especially in projects where there is no donor agency involvement—of ignoring social factors where there are no apparent consequences on the natural environment. It is therefore imperative that the DOE develops separate SIA guidelines similar to its guidelines for EIA. Appropriate amendment to the legislation on EIA is required to provide legal status to SIA. It needs to be established that SIAs are obligatory for socially significant projects even if there are no apparent impacts on the biophysical components of the environment (Momtaz, 2006).

5.5.2 Adequacy of Administrative Arrangement

The ECA '95 confers authority to the DOE to provide guidance to the proponents, review EIAs, and, ultimately, ensure the consideration of social impacts of proposals. However, this organization does not have adequate resources or expertise to enforce ECA '95 fully and effectively and to ensure proper implementation of mitigation measures (see also Chapter 3). It does not have an EIA/SIA database to help project proponents. Also, it does not necessarily share whatever information it has with others. Strengthening DOE and making it more efficient, transparent, and accountable would be a step forward toward effective implementation of SIA in Bangladesh.

5.5.3 Adequacy of SIA Practice

Development partners and international donor agencies play an important role in the inclusion of social issues in the conduct of EIA for the projects that receive their funding. These organizations have institutionalized SIA through numerous guidelines and methodologies. They have developed training programs for their staff engaged in the implementation and review of EIA. Release of funds is often linked to the fulfillment of conditions such as proper consideration of social issues in project development. There is, however, a lack of checks and balances in the conduct of EIA and SIA by private consultants especially for the smaller projects with no donor involvement.

Community consultation constitutes an important ongoing activity in SIA. In recent times, NGOs and donors have developed community consultation guidelines (CEGIS, 2004) and enforced those into the conduct of EIA and SIA (see the examples of EIA and SIA above). SIA review for this study shows that, in most projects, the proponents followed appropriate community consultation methods and the community had the opportunity to contribute to the conduct of SIA (Table 5.7). However, in some cases, this is just a public relations exercise rather than a mechanism that is designed to empower people through greater ownership of and control over the outcome of decision making. Often, community involvement ends when the EIA study is completed (see Chapter 7). Furthermore, community (especially the less-educated people) often does not get the opportunity to comment on draft EIAs. There is yet no provision in the EIA procedure for public display of EISs.

Practice of EIA and consideration of social impacts in the public and private sectors where there is no supervision of donor agencies is a less explored area of environmental management in Bangladesh (see above). Politicization of EIA process and all pervasive corruption have made it nearly impossible to conduct fair EIA or social assessment in these sectors. No in-depth study has been conducted in this area. Research is needed to investigate how EIA and, therefore, SIA are managed in private sector and in projects where there is no donor agency supervision and how these sectors could be incorporated into the mainstream EIA and SIA in Bangladesh (Momtaz, 2005). The quality of SIA would be directly correlated to the quality of the consulting firm conducting the SIA. It is therefore important that DOE develops measures to control the quality of local consultants.

5.5.4 Adequacy of Foundational Measures in the Bangladesh SIA System

As stated earlier, the SIAs prepared in Bangladesh are generally of good quality (Table 5.7). This can be attributed to the guidelines prepared by the DOE, NGOs, and international development partners, and also to the role played by donor organizations in the implementation of EIA. Regular training programs by the donor organizations and the long experience in EIA practice in Bangladesh have resulted in the development of good knowledge base and expertise in EIA and SIA. However, while most social impact statements have done well in the mitigation and monitoring area, which means that most SISs have been good at "monitoring arrangements" and "environmental management plans" (see Table 5.7), actual monitoring of impacts at the post-EIS level has remained the weakest link in the absence of effective supervision of DOE in this area.

Strategic environmental assessment (EIA at policy level), integrated impact assessment (combining all forms of assessment into one study where relevant), and cumulative impact assessment (consideration of cumulative effects while conducting project EIA) are increasingly becoming a norm in the developed world. This is due to the realization that project-level EIA of projects with wider implications may not always reveal the real magnitude of impacts and thereby may lead to inadequate mitigation and monitoring measures.

In Bangladesh, SIAs are basically conducted at project level. LGED, CEGIS, World Bank, and ADB—all these organizations have involvement in projects that are listed by the government as top priority poverty alleviation projects. At the moment, cooperation and coordination among these organizations are an exception rather than the norm. Better coordination among them would help avoid duplication of SIA in those areas where various organizations are involved in some form of project intervention. This would also lead to the development of strategic or policy-level SIA. SEA is yet to have any footing in policy formulation in Bangladesh, let alone its implementation.

5.5.5 Review of EIA and SIA

The final recommendation of this chapter is that a major review of EIA and SIA in Bangladesh is conducted in order to find out how EIAs and SIAs have been implemented since the enactment of ECA 1995; how recommendations of those EIAs and SIAs have been adopted by the proponents; whether and how mitigation measures are implemented; and finally, whether EIAs and SIAs have been able to ensure environmentally and socially sound development projects. There should be specific mention of SIA in the reviewed ECA. This review should also consider developing national guidelines for SIA and advocating the creation of a legal basis for SIA in Bangladesh. This book is an attempt in the same vein.

5.6 CONCLUSION

This chapter reviews the performance of SIA system in Bangladesh and identifies its strengths and weaknesses. It is suggested that necessary amendment is made to the legislation, a thorough review of EIA and SIA process is conducted for necessary reform, and private sector development assessment is brought in line with the mainstream EIA/SIA. Currently, SIAs are conducted on an *ad hoc* basis. The process has huge potential in Bangladesh to make real contribution to ensuring long-term social sustainability as outlined by Vanclay and Esteves (2011). Realization of this potential will, however, depend on how the government and the development agencies address the issues revealed in this study. A coordinated approach to strategic social assessment, supported by a legislative mandate and strengthened lead agency, will improve the ability of Bangladesh to address social issues of development more effectively.

APPENDICES

Appendix 5.1 Reports Reviewed for the Quality of SIA in Bangladesh

Sl. No	Name of projects	Year of approval	Proponents
1	Environmental analysis of Second Road Rehabilitation and Maintenance Project	2000	Ministry of Local Government and Rural Development

Appendix 5.1 Reports Reviewed for the Quality of SIA in Bangladesh—cont'd

Sl. No	Name of projects	Year of approval	Proponents
2	EIA and SIA of Nalka-Hatikamrul-Bonopara New Road	2000	Ministry of Communication
3	EIA of Phulbari Coal Mine Project	2000	Ministry of Energy, Power and Mineral Resources
4	EIA of Gorai River Restoration Project	2001	Ministry of Water Resources
5	EIA of Southeast Area Integrated Water Resources Planning and Management Project	2005	Ministry of Water Resources
6	EIA of detail design of Civil Works Under Road Network and Improvement Project	2007	Ministry of Communication
7	JBIC Eastern Bangladesh Bridge Improvement	2007	Ministry of Communication
8	EIA of Third Karnaphuli Bridge Project	2007	Ministry of Communication
9	EIA of Siddhirganj-Maniknagar 230 kV Transmission Line Project	2008	Ministry of Energy, Power and Mineral Resources
10	EIA of 2 × 150 MW Gas Turbine Power Plant at Siddhirganj	2008	Ministry of Energy, Power and Mineral Resources
11	IEE, EIA, and RAP of Bakhrabad-Siddhirganj Gas Transmission Pipeline Project	2008	Ministry of Energy, Power and Mineral Resources

Appendix 5.2 Coding Protocol for Evaluating SIAs

	Poor	Medium	Good
Executive summary	Summary is not clear. Used technical words without clarification	Intermediate between poor and good	Clearly summarizes the report. Used appropriate terminology for knowledgeable laypersons

(Continued)

Appendix 5.2 Coding Protocol for Evaluating SIAs—cont'd

	Poor	Medium	Good
Description of natural environment	Irrelevant data on social and biophysical environment. Selective omission of existing important projects	Intermediate between poor and good	Detailed description of biophysical, natural, and social environment. Clear identification of existing projects
Description of project	Presentation of partial data on project not covering all phases of project construction and implementation	Intermediate between poor and good	Detailed description of all activities of all phases of the project
Impact identification and prediction methods	Use of very general methods (yes, no, perhaps questions in the questionnaire) without providing adequate justification of answers	Use of simple methods (yes, no, perhaps questions in the questionnaire) but detailed justification of answers. Some important impacts are not clearly explained	Use of proven methodologies for the description and identification of impacts (checklists, matrices, etc.) with full documentation of the expected positive and negative impacts
Mitigation and monitoring	Some mitigation measures have been described	Mitigation measures have been described for the majority of the impacts	Mitigation measures for all identified impacts have been described clearly. Enhancement of positive impacts has also been discussed

Appendix 5.2 Coding Protocol for Evaluating SIAs—cont'd

	Poor	Medium	Good
Alternatives	Some discussions on alternatives (mostly one, i.e., location)	Intermediate between poor and good	Providing analysis of all possible alternatives with respective positive and negative impacts
Public participation	Selective mention of groups in favor of project	Mention participant groups but do not provide details of discussions	Description of involvement of all stakeholders, their issues, how their involvement led to project modification

REFERENCES

Ahammed AKMR, Harvey N. Evaluation of environmental impact assessment procedures and practice in Bangladesh. Impact Assess Project Appraisal 2004;22(1):63–78.

Ahmadvand M, Karami E, Zamani GH, Vanclay F. Evaluating the use of social impact assessment in the context of agricultural development projects in Iran. Environ Impact Assess Rev 2009;29:399–407.

Androulidakis I, Krakassis I. Evaluation of the EIA system performance in Greece, using quality indicators. Environ Impact Assess Rev 2006;26:242–56.

Asia Energy Corporation. Phulbari coal project: environmental impact assessment. Dhaka, Bangladesh: Asian Development Bank; 2006.

Asian Development Bank (ADB). Handbook on social analysis; 2007 [online]. Available from http://www.adb.org/Documents/Handbooks/Social-Analysis/default.asp [Accessed 12 March 2011].

Baines J, Taylor C. Ethical issues and dilemmas. In: Vanclay F, Esteves M, editors. New directions in social impact assessment. Cheltenham, UK: Edward Elgar; 2011. p. 96–116.

Bangladesh Centre for Advance Studies (BCAS). Guide to the environmental conservation act 1995 and rules 1997. Dhaka, Bangladesh: BCAS; 1999.

Burge R. Benefiting from the practice of social impact assessment. Impact Assess Project Appraisal 2003;2(3):225–9.

Canelas L, Almansa P, Merchan M, Cifuentes P. Quality of environmental impact statements in Portugal and Spain. Environ Impact Assess Rev 2005;25:217–25.

Cashmore M, Christophilopoulos E, Cobb D. An evaluation of the quality of environmental impact statements in Thessaloniki, Greece. J Environ Assess Policy Manage 2002;4(4):371–95.

Center for Environmental and Geographic Information Services (CEGIS). Community consultation guidelines. Dhaka: CEGIS; 2004.

Center for Environmental and Geographic Information Services (CEGIS). Social impact assessment/resettlement action plan of Siddhirganj-Maniknagar 230 KV transmission lines project. Dhaka: CEGIS; 2008.

Cottrell A, King D. Disasters and climate change. In: Vanclay F, Esteves M, editors. New directions in social impact assessment. Cheltenham, UK: Edward Elgar; 2011. p. 154–70.

Cox G, Dale A, Morrson T. Social assessment and resource management in Australia. In: Dale A, Taylor N, Lane M, editors. Social assessment in natural resource management institutions. Collingwood, Australia: CSIRO Publishing; 2000. p. 74–92.

Department of Environment (DOE). The Environmental Conservation Rules. Dhaka, Bangladesh: Ministry of Environment and Forest; 1997.

Department of Environment (DOE). Environmental Conservation Amendment Act 2009. Dhaka, Bangladesh: Government of Bangladesh; 2009.

Esteves A, Barclay M-A, Brereton D, Samson D. Enhancing the benefits of projects through local procurement. In: Vanclay F, Esteves M, editors. New directions in social impact assessment. Cheltenham, UK: Edward Elgar; 2011. p. 233–52.

Flood Action Plan (FAP). Guidelines for environmental impact assessment. Dhaka, Bangladesh: FAP; 1992.

Franks D, Brereton D, Moran C. Cumulative social impacts. In: Vanclay F, Esteves M, editors. New directions in social impact assessment. Cheltenham, UK: Edward Elgar; 2011. p. 202–20.

Glasson J, Therivel R, Chadwick A. Introduction to environmental impact assessment. London: Taylor & Francis Group; 2005.

Government of Bangladesh (GOB). Bangladesh National Environmental Policy 1992. Dhaka: GOB; 1992.

IAIA (International Association for Impact Assessment). Social impact assessment international principles; 2003 [online]. Available from http://www.iaia.org [Accessed 22 June 2011].

Interorganizational Committee on Guidelines and Principles (ICGP). Guidelines and principles for social impact assessment. US Dep Commer NOAA Tech Memo NMFS-F/SPO-16; 1994. p. 29.

Lahiri-Dutt K, Ahmad N. Considering gender in social impact assessment. In: Vanclay F, Esteves M, editors. New directions in social impact assessment. Cheltenham, UK: Edward Elgar; 2011. p. 117–37.

Lee N, Colley R. Reviewing the quality of environmental assessments. Occasional paper, Number-24. EIA Centre, UK: University of Manchester; 1992.

LGED (Local Government Engineering Department). Small scale water resources development sector project guidelines for environmental assessment. Dhaka: LGED; 1992.

LGED (Local Government Engineering Department). Environmental assessment guidelines for LGED projects. Dhaka: LGED; 2007.

Louis Berger Group. Road network improvement and maintenance project: environmental impact assessment. Dhaka: Government of Bangladesh; 2007.

MacNaughton G, Hunt P. A human rights-based approach in social impact assessment. In: Vanclay F, Esteves M, editors. New directions in social impact assessment. Cheltenham, UK: Edward Elgar; 2011. p. 355–68.

Ministry of Environment and Forest (MoEF). National Environmental Management Action Plan. Dhaka, Bangladesh: Ministry of Environment and Forest; 1995.

Modak P, Biswas AK. Conducting environmental impact assessment for developing countries. New York: United Nations University Press; 1999.

Momtaz S. Environmental impact assessment in Bangladesh: a critical review. Environ Impact Assess Rev 2002;22:163–79.

Momtaz S. The practice of social impact assessment in a developing country: the case of environmental and social impact assessment of Khulna-Jessore Drainage Rehabilitation Project in Bangladesh. Impact Assess Project Appraisal 2003;21(2):125–32.

Momtaz S. Institutionalizing social impact assessment in Bangladesh resource management: limitations and opportunities. Environ Impact Assess Rev 2005;25:33–45.

Momtaz S. Public participation and community involvement in environmental and social impact assessment in developing countries. Int J Environ Cult Econ Soc Sustain 2006;2(4):89–97.

O'Faircheallaigh C. Social impact assessment and indigenous social development. In: Vanclay F, Esteves M, editors. New directions in social impact assessment. Cheltenham, UK: Edward Elgar; 2011. p. 138–53.

Sandham LA, Pretorius HM. A review of EIA report quality in the North West province of South Africa. Environ Impact Assess Rev 2008;28(4–5):229–40.

SAPROF. JBIC special assistance for project formation (SAPROF) for Eastern Bangladesh bridge improvement project: environmental impact assessment. Dhaka: Japan Bank for International Cooperation; 2007.

Vanclay F, Bronstein D. Environmental and social impact assessment. New York: J Wiley; 1995.

Vanclay F, Esteves AM, editors. New directions in social impact assessment. Cheltenham, UK: Edward Elgar; 2011.

Water Resources Planning Organization (WARPO). Guidelines for environmental assessment of flood control. GOB, Dhaka, Bangladesh: Drainage and Irrigation Projects; 2011.

WB (The World Bank Group). Social analysis sourcebook: incorporating social dimension into Bank-supported projects. Washington DC, USA: WB; 2002.

CHAPTER 6

Evaluating Community Participation in Environmental Impact Assessment

Contents

Evaluating Environmental and Social Impact Assessment in Developing Countries © 2013 Elsevier Inc.
http://dx.doi.org/10.1016/B978-0-12-408129-1.00006-1   All rights reserved.

6.1 INTRODUCTION AND BACKGROUND

In the recent past, the governments in developed and developing countries have provided legislative mandate for community participation (CP) in environmentally and socially significant decisions. While various forms of community involvement have become firmly established in developed countries and community's role in environmental and social impact assessment has been thoroughly explored in recent literature (Burdge and Robertson, 1990; Cox et al., 2000; Vanclay and Bronstein, 1995), it remains a less investigated area of development intervention in developing countries. This chapter examines the status of CP in Bangladesh as highlighted in various guidelines and practiced in project implementation. It describes the legal and organizational frameworks of CP and reports two examples of EIA and SIA with significant community involvement. These two projects are further elaborated in the case study chapter (Chapter 7).

Community consultation and participation in environmental policy making has been enshrined in major environmental documents and legislation. Agenda 21 has recognized CP as a prerequisite for sustainable development (Agenda 21, section 23). The role of CP in the collection of local and indigenous knowledge has also received greater recognition in the sustainable development debate. Principle 10 of the Rio Declaration on Environment and Development recognizes, "environmental issues are best handled with the participation of all concerned citizens at the relevant level."

A number of recent research investigated into CP as a decision-making tool. Irvin and Stansbury (2004) identify the conditions under which CP may be costly and ineffective and the circumstance under which CP can produce effective citizen governance. They argue that CP may not always be an effective decision-making tool. Beder (1999) recognizes the importance of public participation but argues that the whole process may be used as a public relations exercise to manipulate people's opinions and perceptions. Petts (1999) identifies various methods of CP and examines the effectiveness of two methods, namely, advisory committees and citizen's juries, in rural policy in the United Kingdom. Increased involvement of community in decision making may be seen as an inevitable result of globalization and the desire of the mature democracies to place more responsibilities on citizens and less on the state.

In summary, CP has been accepted as a means to achieving sustainability in plans, policies, and projects. Authors have indicated that the debate surrounding participation is no more on its desirability rather on "what type of citizen-participation process is best" (Irvin and Stansbury, 2004, p. 56). There is a need for more investigation into the question—what constitutes effective CP?

Despite the debate on what constitutes CP and whether the terms "involvement," "participation," and "consultation" should be used interchangeably to mean "community participation in environmental decision-making," community involvement has become an integral part of environmental, social, and almost in all forms of impact assessments since NEPA 1969. Developed countries have legislated community involvement in environmental decision making. Similarly, recently developing countries are also making efforts to involve affected people in mitigating impacts and developing management plans. International development partners and local nongovernmental organizations (NGOs) are playing important roles in strengthening CP.

The authors have conducted a thorough analysis of contents of EIA legislation and all major EIA documents (see also Chapters 3–5) to evaluate the status of CP in Bangladesh. An in-depth examination of two of the biggest development projects—the Jamuna Bridge Project and Khulna Jessore Drainage Rehabilitation Project—where CP played a role has also been done. While the content analyses shed lights into the legislative side, the case studies including the ones in Chapter 7 provided an insight into the applied side of relevant legislation and guidelines in the post-EIS stage.

6.2 CONCEPTUALIZING COMMUNITY PARTICIPATION: ORIGIN, DEFINITION, AND EVOLUTION

CP in environmental decision making is about involving people and parties that are likely to be directly or indirectly affected by the decision. These individuals and groups are identified during the scoping phase of an environmental impact assessment usually known as stakeholder identification.

6.2.1 Community Participation in NEPA 1969

Council of Environmental Quality established under the Act has been given the authority to consult various groups and community representatives to exercising its powers. NEPA requires involvement of public in the scoping phase,

> ...There shall be an early and open process for determining the scope of issues to be addressed and for identifying the significant issues related to the proposed action. Significant issues are to be identified and insignificant issues are to be eliminated from the detailed study, through consultation with all interested parties.
>
> **NEPA '69, p. 7**

Guidelines and Principles for Social Impact Assessment Prepared by the Interorganizational Committee on Guidelines and Principles for Social Impact Assessment (1994) is the first SIA document after NEPA from the United States that clearly outlines the links between environmental and social impact assessment. It identifies development of a plan to involve all potentially affected people as the first step in SIA process.

> This requires identifying and working with all potentially affected groups starting at the very beginning of planning for the proposed action. Groups affected by proposed actions include those who live nearby; those who will hear, smell or see a development; those who are forced to relocate because of a project; and those who have interest in a new project or policy change but may not live in proximity (p. 11).

6.2.2 Community Participation in Agenda 21

Agenda 21 is one of the major documents that came out of United Nations' Rio Summit on Environment and Development in 1992. It is a comprehensive plan of action, recommended by UN summit to be taken globally, nationally, and locally by organizations of the United Nations System, governments, and major groups in every area in which human impacts on the environment.

Agenda 21 identifies broad CP as one of the prerequisites for sustainable development. Chapter 23 of Agenda 21 is about "Strengthening the role of major groups." The "Preamble" says

> ...in the more specific context of environment and development, the need for new forms of participation has emerged. This includes the need of individuals, groups and organizations to participate in environmental impact assessment procedures and to know about and participate in decisions, particularly those which potentially affect the communities in which they live and work.
>
> **Agenda 21, Preamble 1992**

6.2.3 NOAA Guidelines

Due to the growing support for stakeholder participation early in the policy development phase, the National Oceanic and Atmospheric Administration provided administrative issuance (NOAA 216-6, 2011) *Environmental*

Review Procedure for Implementing National Environmental Policy Act Public Involvement. It explains various aspects of NEPA and NOAA's compliance with the legislation. It emphasized,

> Public involvement is essential to implementing NEPA. Public involvement helps the agency understand the concerns of the public regarding the proposed action and its environmental impacts, identify controversies, and obtain the necessary information for conducting the environmental analysis. RPMs must make every effort to encourage the participation of affected Federal, state, and local agencies, affected Indian tribes, and other interested persons throughout the development of a proposed action and to ensure that public concerns are adequately considered in NOAA's environmental analyses of a proposed action and in its decision-making process regarding that action NOAA Administrative Issuance.
>
> **NOAA 216-6 (2011)**

6.2.4 Public Participation in International Association for Impact Assessment Guidelines

International Association for Impact Assessment (IAIA) has defined public participation as "the involvement of individuals and groups that are positively or negatively affected by a proposed intervention (e.g., a project, a program, a plan, a policy) subject to a decision-making process or are interested in it" (André et al., 2006, p. 1). The basic principles as indicated by the documents are adapted to the context, informative and proactive, adaptive and communicative, inclusive and educative, and cooperative and imputable.

The purpose of stakeholder involvement (in other words, community involvement) in policy making has been described variously by authors. A quick review of recent articles on CP-related topics reveals a long list of purposes that community consultation or involvement intend to fulfill. The purposes as indicated in recent literature are collection of information, building support, taking account of values, giving community the ability to influence the outcome of decision or empowerment, avoiding litigation, strengthening the democratic fabric of society, etc. (Lockie et al., 1999; Momtaz and Gladstone, 2008). This is true that the nature of CP in environmental decision making has been shaped in Western democracies by people's desire to take greater responsibility and active role in natural resource management rather than being passive beneficiaries or recipient of negative impacts of decisions. Greater participation of community is also seen as an important aspect of maturing democracies. However, there is a debate that multiple purposes are listed often without differentiating the meaning and without knowing if all the listed purposes and objectives are likely to be fulfilled in each case (O'Faircheallaigh, 2010). Various forms of CP and involvement are practiced in developing countries. The nature and quality of participation is often determined and shaped by the conviction of the governments, influence of the civil society, and role of the development partners. In most cases, the form of participation practiced in Western democracies may not be appropriate for developing countries. While some Asian developing countries, such as India, Indonesia, The Philippines, Taiwan, and Hong Kong, are in the forefront of EIA practice, African and Latin American countries are still in the process of catching up.

6.3 COMMUNITY PARTICIPATION IN BANGLADESH

6.3.1 Legal Framework of Community Participation

Despite being the legal foundation of EIA in Bangladesh, neither ECA '95 nor Environmental Conservation Rules 1997 say much about CP. This was probably due to the fact that defining detailed procedure of CP in EIA was not within the scope of these documents and the legislators left community consultation aspects of EIA to be dealt with within the guidelines drafted by the DOE and others. As stated earlier, Department of Environment (DOE) is the government agency with the authority to enforce ECA '95 and approve EIAs conducted for projects. CP has been indicated in DOE guidelines, as described later in this chapter, as an important component in EIA process and shown to be firmly established in the project cycle.

6.3.2 Community Consultation in EIA/SIA Guidelines

6.3.2.1 Flood Plan Co-Ordination Organization Guidelines for EIA

These are the first EIA guidelines in Bangladesh published in 1992 for flood control, irrigation, and water management projects (Flood Action Plan or FAP). The guidelines were updated in 2001 with greater emphasis on people's participation (Water Resources Planning Organization (WARPO), 2011). The document acknowledges that active participation of stakeholders is mandatory in water resources planning and management and must therefore form an integral part of all environmental assessments. The guidelines clearly outline steps to involving people in initial environmental examination (required for small-scale projects with no significant environmental impacts) and in environmental impact assessment.

6.3.2.2 Guidelines for Environmental Assessment 1992 of Local Government Engineering Department

Local Government Engineering Department (LGED), under the Ministry of Local Government and Rural Development, has the responsibility for small-scale water resources development projects. LGED has recently revised its environmental assessment guidelines for LGED projects. The guiding principle behind formulation of the Guidelines is to safeguard the physical, biological, and socioeconomic environments during project preparation, implementation, and operation. The document has clearly set out CP in a conceptual framework (Figure 6.1).

6.3.2.3 DOE's EIA Guidelines for Industries

As indicated earlier, DOE guidelines require that the proponent will incorporate EIA into project planning and maintain liaison with the DOE, concerned departments, local people, and NGOs (DOE, 1997). The document suggests that EIA should involve community in the decision-making process as people are the ultimate recipient of economic benefits and environmental

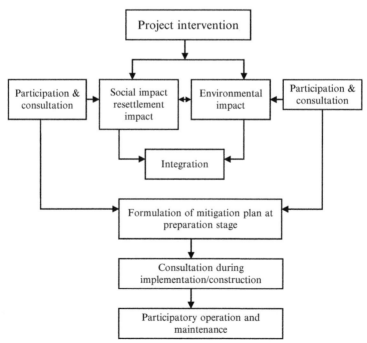

Figure 6.1 Conceptual framework for participation and consultation in LGED. *Source: Prepared by the authors from LGED (2007, p. 4–8)*

damages. For effective and meaningful community involvement, "it is necessary to communicate with as many people as possible, as early as possible, and through as many different ways as possible" (DOE, 1997, p. 34).

6.3.2.4 Guidelines for Participatory Water Management by Ministry of Water Resources

Bangladesh government has initiated discussions to provide a legal framework to people's participation in water sector. With this objective in mind, the Ministry of Water Resources released these guidelines to ensure CP in water resource management programs (Ministry of Water Resources, 2001). It proposes to establish Water Management Organizations consisting of local stakeholders and project-affected people at various administrative and geographic levels.

6.3.2.5 Community Participation in CEGIS Documents

Centre for Environmental and Geographic Information Services (CEGIS) is one of the organizations that has placed significant emphasis on stakeholder participation and used extensive community consultation and participation in all EIA and SIA conducted by it. It is probably the only agency that conducted strategic level community consultation. It carried out consultations in 2004 for the proposed *haor* (wetland) rehabilitation projects by the Bangladesh Water Development Board in 2004-2005.

6.3.2.6 Community Consultation in World Bank Projects

CP is a part of formal environmental assessment of World Bank-funded projects since 1989. The WB documents have emphasized the need for consulting the poor and vulnerable, especially for the projects with significant environmental impacts (Category A) to ensure project sustainability.

6.3.2.7 Community Participation in Asian Development Bank Projects

Asian Development Bank (ADB) has emphasized the need for further strengthening community consultation and participation (C&P) (ADB, 2006). For the purpose of its development and poverty alleviation projects for developing countries, ADB has defined community consultation and participation as "a process through which stakeholders influence and share control over development initiatives, and the decisions and resources that affect them." Some important purposes of C&P as identified by ADB are information sharing, gaining inputs from stakeholders, and collaborative decisions (ADB, 2006).

As can be seen from the above discussion, CP has been treated with much emphasis in various guidelines and documents. Different organizations have developed different methods of consulting and involving affected people and stakeholders in order to fulfill the obligations as outlined in these documents. However, the question is whether this ethos of good guidance for public participation in the form of guidelines and documents is translated into reality. Answer to this question can be found in the case studies (Chapter 7).

6.4 COMMUNITY PARTICIPATION IN EIA AND SIA: TWO CASE STUDIES

These two case studies, along with another project, were also the subject of detailed investigation for this book to determine effectiveness of EIA system in the post-EIS phase. In the following sections, we have focused mostly on the community consultation aspect of these two projects. Detailed project descriptions and social and environmental impacts are provided in Chapter 7.

6.4.1 Case Study 1: Community Participation in Jamuna Bridge Construction

6.4.1.1 Project Description
The project was the biggest infrastructure development in Bangladesh with an estimated cost of $700 million to construct a 4.8-km-long bridge over the river Jamuna.

6.4.1.2 Social Impacts
The major social impacts include involuntary relocation of several thousand households and influx of external workforce in the area creating social and economic issues. In response to these revelations, the authority commissioned a separate compensation and resettlement planning study to define and identify project-affected people and develop Resettlement Action Plan (RAP).

6.4.1.3 Community Consultation
Substantial public consultation and involvement took place during the EIA and other planning studies, especially during the socioeconomic/population surveys at the scoping phase. Consultation included: (a) visits to the project sites and discussions with the project-affected people; (b) formulation of

village committees for efficient liaison and grievance accounting procedure with the participation of representatives of PAPs, social leaders, village elders, interested NGOs, and members of the local organizations; (c) noting and analysis of responses; and (d) use of the ideas in the preparation of the RAP. Project-affected people were categorized according to the extent of loss incurred by the proposed development and provisions and entitlements were determined accordingly through the development of a resettlement policy matrix (RPM). A section of the RPM has been shown in Table 6.1. This matrix was based on wide community consultation and the information collected through surveys.

Table 6.1 Resettlement policy matrix for Jamuna Bridge Railway Link Project

Description of category	Provisions and entitlement	Comments
Land owners losing only a portion of their agricultural land. Left with residue of two or more acres with H/H of six persons. A fraction thereof for each additional person	Cash compensation for land or as for categories 2 and 3 below	Land owning categories who are full-time farmers (and not landholders) are extremely vulnerable to land acquisition or significant magnitude as they rarely have secondary sources of income
Owners of agricultural land with up to six H/H members who will be left with less than two acres, or proportionately more for each additional person	Cash compensation for land acquired and additional cash to purchase similar land elsewhere	PAPs to be informed that they can look for agricultural land for sale, individually or in groups. Authority will assist them to purchase the land
Owners losing all their land	As for category 2 above	As for 1 and 2 above
Sharecroppers with no land	Receives cash compensation for homestead and house Receives priority offer to lease land along approach roads Proposals for vocational training to be prepared with help of NGOs	

Table 6.1 Resettlement policy matrix for Jamuna Bridge Railway Link Project—cont'd

Description of category	Provisions and entitlement	Comments
Shopkeepers, kiosk and stall owners	Cash compensation for moving business	
Homestead owners losing house and land on which it stands	Can get compensation for (a) land, (b) house, both at replacement value	
Squatters and farm laborers	Receives cash compensation for homestead and house Receives priority offer to lease land along approach roads Proposals for vocational training to be prepared with the help of NGOs	
Weavers and other industries and artisans	Cash grant Vocational training Some land will be made available by the authority on both sides and at optional settlement sites	
People adversely affected by bridge, i.e., change in water levels upstream or downstream, or in unforeseeable ways	Cash compensation and assistance to reestablish workshops at new sites selected by them	Legal agreement between the authority and the government to be drawn up to protect interests of all persons who may be adversely affected by the bridge and allied civil works
Persons whose property has already been acquired for the project	Assistance to mitigate impacts, e.g., through provision of pumps	

Source: Jamuna Bridge EIA.

6.4.2 Case Study 2: CP in Khulna Jessore Drainage Rehabilitation Project

6.4.2.1 Project Description

The project area comprises numerous tidal rivers and creeks, which provide a drainage network to a system of embanked hydrological units or polders. The construction of embankments in the 1960s under the coastal embankment project greatly reduced the volumes of water entering and leaving the project area during the tidal cycles. This resulted in gradual siltation of the drainage networks and drainage congestion. Waterlogging took away 20% of land from agricultural production and forced 78% of the households to migrate.

In 1995, the government of Bangladesh initiated the Khulna Jessore Drainage Rehabilitation Project to find more permanent relief to the suffering of the local people. CEGIS was contracted to conduct an independent EIA and SIA study.

6.4.2.2 Social and Economic Impacts
All options of interventions were likely to have positive and negative socioeconomic impacts in the form of impacts on occupation, income, capture fisheries, the possibility of homestead inundation, health, education, and women's activities (CEGIS, 1998).

6.4.2.3 Community Consultation
The study started with intensive consultation with local people and NGOs. Raid Rural Appraisal, conducted in 60 different spots, involved extensive discussions with people belonging to various socioeconomic strata and played an important part in identifying project-affected people. Separate sessions were held with various socioeconomic groups to facilitate maximum participation. One of the results of the public participation was the realization by the study team that the general public had an option to offer for addressing the waterlogging problem in the project area which was better to the options resulting from the feasibility studies. This community-identified option was treated equally at the "consideration of alternatives" phase in the EIA process.

One important outcome of the consultation process was that in the recommendations of SIA/EIA, CEGIS proposed the formation of water management associations (WMAs) in various parts of the project areas consisting of local community members. CEGIS recommended that the WMAs would have the responsibility to supervise and monitor the project area during and after the intervention.

However, detailed investigation at the post-EIS phase (see Chapter 7) revealed that the proponent did not take the community-proposed option very seriously and that community involvement at the implementation and monitoring phase of the project was nominal.

6.5 LESSONS LEARNT FROM BANGLADESH EXPERIENCE

6.5.1 CP is Well Established in Guidelines

As yet, there are no legislative directives for CP in Bangladesh. ECA '95 does not provide any instructions on CP. However, as demonstrated earlier, it is well established in guidelines developed by various lead organizations. International donor agencies and development partners (e.g., USAID, CARE, World Bank, Asian Development Bank) played significant role in the formulation and implementation of community involvement processes in EIA and SIA.

Despite the fact that provision for CP is well embedded in decision-making process, a solid legislative mandate is necessary. This will allow the government and the environmental agencies to bring all community involvement processes within the legal framework. In the mean time, the DOE should ensure strong supervision in the implementation of community consultation process in the conduct of EIA and SIA, and international donors and development partners should continue to provide guidance. This should make up for the lack of legislative mandate for CP.

6.5.2 Community Participation Played a Role in Decision Making

In the case studies, people's participation played an important role in the identification, evaluation, and assessment of impacts. Community opinions also influenced outcomes of decision making in terms of preferred options, development of mitigation measures, and environmental management methods. In the end, CP paved the way for enhancement of positive impacts and mitigation of negative effects. It is not enough for developing countries to have legislation and guidelines for CP which most Asian countries have; more important, perhaps, is to put mechanisms in place so as to make participation effective. The case studies (Chapter 7) will reveal that there is scope for further improvement in community involvement, especially, in the post-EIS project implementation, environmental management and monitoring, and mitigation.

6.5.3 Community Participation and Involvement Created Sense of Ownership and Resulted in Better Cooperation

One of the important aspects of CP is collecting and sharing information and gaining understanding of the community values and aspirations (Mitchell, 2002). Public involvement is also about a greater recognition of the fact that local people have wealth of local knowledge that is acquired through trial and error in the field over a long period of time. This knowledge can be integrated into the scientific knowledge of EIA and SIA and help develop a management plan that is appropriate for each situation. Furthermore, community involvement is also about seeking opinion of the community on various aspects of project impact. Formation of community groups for the purpose of ongoing communication created a sense of ownership in the community and empowered them with the ability to influence the outcomes of decisions that had the potential to transform the community for good.

6.5.4 NGOs Can Play an Important Role in Organizing Local People for Involvement in Decision-Making Process

Because of their grass roots base, NGOs have the ability to organize the project-affected people and mobilize local resources. NGOs have established themselves as alternative development partners in Bangladesh. So far, they have been quite successful in bringing people above the poverty line through various cooperative programs. The same spirit of cooperation can be utilized in ensuring sustainability of development projects. Numerous NGOs are working in Bangladesh. Many of them are local branches of international donor agencies like Oxfam, CARE, and CARITAS. Some local microcredit organizations like Grameen Bank and BRAC have emerged as significant powers in mobilizing local people and resources (Holcome, 1995; Momtaz, 2003) through their training and income–generating activities with the rural poor. These organizations operate very closely with local people as they provide credit and other assistance to groups and individuals. Developing countries can utilize this local force in order to ensure effective participation in project development and implementation. There is hardly any reason for governments and proponents not to involve people that are already well organized through microcredit movement and have the potential to make positive contributions to decision making. This can only enhance the benefits of projects through establishment of trust, sharing of knowledge, and empowerment of local people.

6.5.5 Donor Agencies Should Be Given a Supervisory Role

As stated earlier, donor agencies and development partners helped develop CP framework in developing countries. They have significant financial involvement in all major development activities. EIA and SIA are mandatory in major interventions funded by these organizations as is CP. Governments should allow these organizations to maintain their supervisory role in the implementation of EIA and SIA of large projects to ensure their sustainability.

6.6 CONCLUSION

Despite the absence of specific legislation in Bangladesh, importance of community consultation has been recognized in all major documents. Donor agencies and local NGOs are playing catalyst role in the implementation of CP. In major development projects, community consultation has led to consultative processes that started in the policy formulation phase and continued into decision making. Through community consultation managers provided information to project-affected people and stakeholders, collected information, and involved community to secure a better decision. This democratic process was instrumental in gaining public support. However, questions remain about the role of CP in the post-EIS project implementation phase. Chapter 7 will further look into this matter.

REFERENCES

Agenda 21 1992. [online] Available from http://www.un.org/esa/dsd/agenda21/ [accessed 20.10.11].

André P, Enserink B, Connor D, Croal P. Public Participation International Best Practice Principles. Special Publication Series No. 4. International Association for Impact Assessment, Fargo, USA; 2006.

Asian Development Bank (ADB). Strengthening participation for development results. ADB, Manila: A Staff Guide to Consultation and Participation; 2006.

Beder S. Public participation or public relations? In: Brian martin, editor. Technology and public participation. Wollongong, Australia: University of Wollongong; 1999. p. 169–92.

Burdge RJ, Robertson RA. Social impact assessment and the public involvement process. Environ Impact Assess Rev 1990;10:81–90.

Centre for Environmental Geographical Information Services (CEGIS) . Environmental and social impact assessment of Khulna-Jessore Drainage Rehabilitation Project. Dhaka, Bangladesh: CEGIS; 1998.

Cox G, Dale A, Morrson T. Social assessment and resource management in Australia. In: Dale A, Taylor N, Lane M, editors. Social assessment in natural resource management institutions. Collingwood, Australia: CSIRO Publishing; 2000. p. 74–92.

Department of Environment (DOE) . EIA Guidelines for Industries, Ministry of Environment and Forest. Dhaka: Government of Bangladesh; 1997.

Holcome S. Managing to empower: the Grameen Bank's experience of poverty alleviation. London: ZED Books; 1995.

Interorganizational Committee on Guidelines and Principles (ICGP). Guidelines and principles for social impact assessment, US. Dep. Commer., NOAA Tech. Memo.NMFS-F/SPO-16; 1994, p. 29.

Irvin RA, Stansbury J. Citizen participation in decision making: is it worth the effort? Public Admin Rev 2004;64(1):55–65.

Lockie S, Momtaz S, Taylor B. Meaning and construction of social impacts: water infrastructure development in Australia's Gladstone/Calliope region. Rural Society 1999;9 (3):529–42.

Ministry of Water Resources. Online resources, http://www.mowr.gov.bd/; 2001 [accessed 20 July 2011].

Mitchell B. Resource and environmental management. New York: Prentice Hall; 2002.

Momtaz S. The practice of social impact assessment in a developing country: the case of environmental and social impact assessment of Khulna-Jessore Drainage Rehabilitation Project in Bangladesh. Impact Assess Project Appraisal 2003;21(2):125–32.

Momtaz S, Gladstone W. Ban on commercial fishing in the estuarine waters of New South Wales, Australia: community consultations and social impact. Environ Impact Assess Rev 2008;28(2–3):214–25.

National Environmental Policy Act (NEPA). [online] Available from: http://ceq.hss.doe. gov/index.html; 1969 [accessed 01.12.11].

National Oceanic and Atmospheric Administration (NOAA). Environmental Review Procedure for Implementing National Environmental Policy Act Public Involvement, NOAA administrative issuance (NAO 216-6). [online] Available from http://www. noaa.gov/; 2011 [accessed 12.11.11].

O'Faircheallaigh C. Public participation and environmental impact assessment: purposes, implications, and lessons for public policy making. Environ Impact Assess Rev 2010;30:19–27.

Petts J. Introduction to environmental impact assessment in practice: fulfilled potential or wasted opportunity? In: Petts J, editor. Handbook of environmental impact assessment. London: Blackwell Science; 1999. p. 3–9.

Vanclay F, Bronstein D. Environmental and social impact assessment. New York: John Wiley; 1995.

Water Resources Planning Organization (WARPO) . Guidelines for environmental assessment of flood control. GOB, Dhaka, Bangladesh: Drainage and Irrigation Projects; 2011.

CHAPTER 7

Implementation of Mitigation Measures: EIA Practice at Post-EIS Stage

Contents

Evaluating Environmental and Social Impact Assessment in Developing Countries © 2013 Elsevier Inc.
http://dx.doi.org/10.1016/B978-0-12-408129-1.00007-3

7.1 INTRODUCTION

Implementation of mitigation measures along with management activities, such as monitoring and community participation, is an integral part of an effective EIA system. It is argued that adequate institutional arrangements and good quality of EISs alone cannot guarantee the protection of the environment, the ultimate aim of EIA. It is important that the mitigation measures are also effectively implemented. Effective implementation of mitigation measures arise when all proposed mitigation measures are completely and adequately implemented (Wood, 1995). It is the effective implementation of mitigation measures at the post–EIS stage of EIA process that can address the predicted environmental and social impacts of the projects (Cashmore et al., 2004). The aim of this chapter is to present findings on the status of mitigation measures in Bangladesh.

In this chapter, three projects are investigated in order to understand the implementation of proposed mitigation measures. Section 7.2 provides project background, environmental impacts, proposed mitigation measures, and findings on the implementation of mitigation measures in the Khulna-Jessore drainage rehabilitation project (KJDRP). Section 7.3 outlines project background, environmental impacts, proposed mitigation measures of Jamuna multipurpose bridge project (JMBP), and findings on the implementation of mitigation measures of the project. Section 7.4 presents project description, potential environmental impacts, proposed mitigation measures, and findings on the implementation of mitigation measures of Meghnaghat Power Plant project (MPPP). Section 7.5 provides summary information on the status of the mitigation measures of three projects and the role of DOE. This is followed by a chapter summary.

Three projects were selected based on development sector, proponent type, size of the project, year of implementation, and the possibility of available data. All three projects went through full EIA processes. All three projects were implemented prior to 2004. Data were collected from documents, direct observation and site visits, and interviews of key informants. Finally, a judgment about the extent of the mitigation measures implemented was made based on the available data. It is important to note that only major mitigation measures for the significant impacts have been investigated given the large size of the projects and limited resources.

7.2 CASE STUDY I: KHULNA-JESSORE DRAINAGE REHABILITATION PROJECT

7.2.1 Location of the Project and Environment of Project Area

The project is located in the southwestern part of Bangladesh (Map 7.1). The area is crisscrossed by hundreds of rivers, tributaries, and canals. Fishing and agriculture are the main sources of income for most people in this area. The region is rich in biodiversity with hundreds of species of birds, fish, and vegetation. The region is influenced by the tidal water of the Bay of Bengal located to the south of the region. The soil in this region is still forming through sedimentation carried by tidal water. The region is only 1–3 m above the average sea level.

7.2.2 Project Background

The objective of this project was to solve drainage congestion problems and improve agriculture in the project area. The drainage congestion was a consequence of the implementation of a project in the 1960s. Prior to the 1960s, this region was regularly inundated by high tides with saline water that restricted the development of agricultural activities in *beels*.[1] The Bangladesh Water Development Board (BWDB) constructed a series of regulators[2]

---

[1] *Beels* are large surface water bodies that accumulate surface runoff water through internal drainage channels. These depressions are mostly topographic lows produced by erosions and are seen all over Bangladesh. *Beels* are small saucer-like depressions of a marshy character. Many of the *beels* are dried up during the winter, but during the rainy season they expand into broad and shallow sheets of water, which may be described as freshwater lagoons (Chakraborty, n.d.).

[2] Regulators are kind of structures built in the open waterways (rivers and big canals) to control water getting in and getting out.

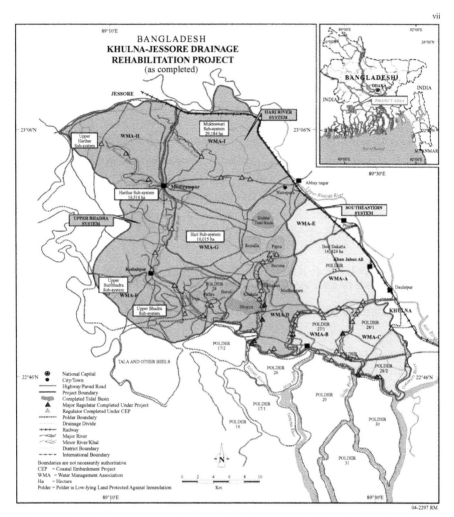

Map 7.1 Location of KJDRP and area under the project. (For color version of this figure, the reader is referred to the online version of this chapter.) *Source: SMEC International, 2002.*

across the rivers in the project area to control the intrusion and the release of tidal water. The BWDB also constructed a series of polders[3] to control the inflow and outflow of tidal water.

---

[3] Polders are low-lying tract of land enclosed by embankments (barriers) known as dikes, that form an artificial hydrological entity, meaning it has no connection with outside water other than through manually operated devices (Wikimedia Foundation, 2011).

The project, immediately after its implementation, created an opportunity for agricultural activities; however, this benefit did not last long. The installation of regulators across the rivers and canals interrupted the natural flow of tidal water. Water could not get in easily in the project area and spread out over the *beels*. Also, the creation of the polders greatly simplified the existing drainage network, comprising a large number of tidal rivers and canals of all sizes. Therefore, heavy siltation occurred on the river beds over the years and caused drainage congestion. As a result, during the rainy season, heavy rain water could not pass easily and a vast amount of land remained waterlogged all the year round. To solve this congestion problem, an intervention was needed. The BWDB, in response, planned to rehabilitate the drainage system of the area through dredging of the rivers, repairing, and installing regulators. This intervention was known as the KJDRP. The implementation of the project began in 1998 and ended in 2002.

7.2.3 Potential Environmental and Social Impacts

The EIA team, under the leadership of the Centre for Environment and Geographic Information System (CEGIS), carried out an Environmental and Social Impact Assessment with the objective of planning an environmentally and socially sound project (GOB, 1998). The EIA team conducted extensive consultations with all of the stakeholders using appropriate methods such as a Participatory Rural Appraisal (PRA) and Focus Group Discussions to identify potential, social, and environmental impacts.

During consultation with the local community, the EIA team came to know that the traditional engineering design of the project (installation of regulators and repairing of the existing ones) would have significant environmental and social impacts, as occurred in past projects implemented in the 1960s. Therefore, the local community proposed to implement an innovative option alternative to the regulator option called the Tidal Basin Management (TBM) or Tidal River Management (TRM[4]). This is an application of a local method called *jower-vata khelano* (free play of tidal water flow) to the rivers and *beels*. This process allows tidal flow into the *beels* of basins and releases the tidal flow daily back to the river. Based on the local community's demand and potential environmental advantage, the TRM option was also incorporated into the project design. The EIA team identified that the TRM option also had some

---

[4] The TRM and TBM will be used interchangeably in this study. This popular concept was developed based on indigenous water management practice developed over generations. The concept later entered into the lexicon of water experts as TRM or TBM.

impacts such as temporary loss of livelihood of farmers or agricultural production. Table 7.1 presents the major environmental and social impacts of KJDRP identified by the EIA team.

7.2.4 Proposed Mitigation Measures

In order to address the impacts, the EIA team recommended a set of mitigation measures (Table 7.2) and monitoring program.

7.2.5 The Status of Implemented Mitigation Measures

7.2.5.1 Implementation of TRM Option

The BWDB implemented the project focusing largely on the construction and rehabilitation of regulators. However, it paid little attention to the proper implementation of the TRM option, a preferred option of the local community. The total project area was divided into four basins (clusters) of *beels*. It was expected that in each of the four basins, *beels* would be brought rotationally under the TRM option. This would make the river navigable and raise the land level of the *beels*. However, the TRM option was implemented only in the Khuksi *beel* and Kaderia *beel* of two separate basins. At present, the TRM of Kaderia *beel* is not under operation and there is no further initiative to bring other *beels* under TRM.

Despite the partial implementation, the TRM option shows better outcomes than the regulator option (Photo 7.1). Discussions with local people and project site visit reveal that the land level of the *beels* under TRM option was raised after a few years owing to sediments carried by the tidal water into the *beels*. The raising of the land level enabled the local farmers to cultivate crops throughout the year.

On the other hand, where regulator options were implemented (repairing the existing ones or construction of new ones), water flows were limited in the rivers. This came about since tidal water could not easily go through the regulators (Photo 7.2). The local community is experiencing the same environmental problem as it was before the implementation of the project.

7.2.5.2 Implementation of Regulator Option

The EIA study has found that the regulator option would have significant impact on fisheries in the project area. To mitigate the loss of open-water fish production, the EIA report proposed for release of fingerlings every year in *beels* of the project area. This is particularly so where regulatory option is in operation. The study also recommended that the regulators should be operated in a fish-friendly way. With this in mind, EIA report proposed

Table 7.1 Major social and environmental impacts of KJDRP

Important environmental issues	Potential significant impacts
1. Loss of open-water fish habitat	a. Reduction of fish habitat for spawning, breeding, and replenishment. b. Discontinuation of migratory routes under the regulatory options and blockage of natural recharge of nutrients and microorganisms.
2. Disruption of agricultural production	Huge amount of dredged material (50,000–400,000 m^3) will be produced from dredging and excavation of rivers and canals to bring the surface water system under reference conditions. Unplanned dumping of dredged materials will create problems in densely populated areas and will affect agricultural land, settlement areas, and communication systems.
3. Quality of water during dredging	During dredging, water quality in the river will deteriorate and, thereby, water users and the habitat in the rivers will be affected.
4. Loss of aquatic habitat for plants and wildlife	There will be qualitative changes in habitat under the proposed project. Area of habitat for birds and other species will be reduced.
5. Possible death of Bhadra River and Hamkura River	Installation of new regulators in the Hamkura River and upper Bhadra River may affect the rivers, and the rivers may die in the long run.
6. Loss of brackish water for shrimp culture	Due to nonavailability of brackish water under regulator options, shrimp ponds will disappear. Consequently, there will be significant loss of employment, production, and income associated with shrimp farming.
7. Loss of farmers' livelihood in the TRM option area	During the TRM, agricultural production will be disrupted for the time being (for 1 or 2 years) until the level of land of the *beels* is raised.
8. Land acquisition	A vast amount of land will have to be acquired for construction of regulators. Many people will lose their agricultural and homestead land and, consequently, their livelihood income will be affected. The amount of land to be acquired will be less in the case of regulator options than the tidal basin option.

Compiled by the authors from Government of Bangladesh (GOB) (1998).

Table 7.2 Recommended mitigation measures for KJDRP

Proposed major mitigation measures	Mitigation actions
1. Protection of open-water fish habitat from loss	a. The extreme loss of habitat under the regulator options can be mitigated; keeping more water and releasing fingerlings every year in wetland project area are required. b. Operating regulators in a fish-friendly way to facilitate fish migration.
2. Management of disposal of dredged materials and disruption of agriculture	Disposal material will be spread uniformly over agricultural land at a maximum depth of 35 cm. These materials can also be used for homestead development.
3. Protection of water quality during dredging	Water quality will be maintained particularly when dredging will be conducted during dry season. During the monsoon, dredging will be stopped for some time for the benefit of breeding fishes.
4. Protection of aquatic habitat for plants and wildlife	Under the tidal basin option, the basins will be maintained as sanctuaries for aquatic plants, fish, and wildlife.
5. Prevention of Bhadra and Hamkura River from dying	Creating a tidal basin in the Singha Beel, this will keep the Bhadra River and lower part of the Hamkura River at reference conditions.
6. Protection of shrimp culture in the project area	Extension services should be provided for encouraging fresh water fish culture. Prospects of limited intake of brackish water through small inlets from rivers to the eastern side of the project area should be explored to facilitate some shrimp farming.
7. Compensation for land acquisition	a. Adequate compensation must be paid to all land owners, including material and technical help. b. Vulnerable Group Development (VGD) program should be introduced for affected small and marginal farmers.
8. Protection of farmers' livelihood in the TRM area	Financial support will be provided for the affected farmers till the land becomes fully cultivable.
9. Plantation in the project area to compensate the loss of vegetation	An enhancement plan will include plantation of selected plants including timber, fruits, and fuel plants.

Compiled by the authors from Government of Bangladesh (1998).

Photo 7.1 Benefits of implementation of TRM option. N.B. Water releases from the *beels* without any interruption. (For color version of this figure, the reader is referred to the online version of this chapter.) *Source: Photo by Kabir from project site visit.*

Photo 7.2 Regulators without fish-friendly structure. N.B. Picture shows vertical slots which are not built or operated in a fish-friendly way. Fish cannot easily move through these slots. (For color version of this figure, the reader is referred to the online version of this chapter.) *Source: Photo by Kabir from site visit.*

for the construction of vertical slots in the regulators in order to facilitate fish migration (GOB, 1998). In order to run this environmental management activity successfully, EIA report also recommended training program for the affected community members.

However, the site visits and interviews with key informants of the local community revealed that the project proponent installed vertical structure at a very limited scale to facilitate fish migration. It was observed that in the Shibnagar and Modhukhali regulators, proponents installed vertical slots (Photo 7.2), but these were not built or operated in a fish–friendly way.

Photo 7.3 Reduction of fish stock in rivers in the project area. N.B. A few fishermen are fishing in the river since the fish stock has reduced due to inadequate implementation of mitigation actions to protect fishery from loss. (For color version of this figure, the reader is referred to the online version of this chapter.) *Source: Photo by Kabir team from site visit.*

Similarly, in the Babadaha and Kaderia *beels*, proponent did not build any fish passes.

The informants explained that the proponent made little attempt to construct fish passes and their operation. The proponent only rehabilitated the regulators by removing its mechanical defects and making it operational. Many of the riverine fish species migrate considerable distances upstream to spawn and travel back. Since fish passes were not built in many of the regulators, this interfered with fish spawning migrations from downstream rivers to floodplains and *beels* in the upstream. Because of the inadequate implementation of the proposed mitigation measure, the loss of fish production could not be offset.

It was not possible to measure the loss of fish stock quantitatively due to the nonavailability of records in the project office or elsewhere. Information from the local informants and other evidence suggest that the inadequate implementation of proposed mitigation actions was largely responsible for the reduction in fish production within the project area (Photo 7.3).

7.2.5.3 Management of Disposal of Dredged Materials

Furthermore, proponent did not adequately implement mitigation actions related to the disposal of dredged soils. Site visit revealed that most dredged and excavated soils were piled up on both sides of the rivers (Photo 7.4). The EIS recommended that the excavated soils needed to be evenly distributed

Photo 7.4 Piled-up dredged soil disrupts agriculture in project area. N.B. Pictures show that excavated and dredged soil is piled up along both sides of the rivers and occupied cultivable land. (For color version of this figure, the reader is referred to the online version of this chapter.) *Source: Photo by Kabir from site visit.*

to the agricultural land nearby because the quality of soil was good for agriculture. However, the proponent did not take adequate initiative to distribute the dredged soils. The undisposed dredged soil on the river banks caused the loss of private agricultural land. This has also affected agricultural production.

7.2.5.4 Protection of Bhadra and Hamkuri Rivers from Dying

The proponent did not implement mitigation action related to the protection of Bhadra River and Hamkuri River from dying. The EIA team predicted that the implementation of the TRM option, particularly in the Singha *beel*, would keep the upper Bhadra River and lower part of the Hamkuri River navigable at reference conditions. Since the proponent did not introduce TBM option in Singha *beel*, the upper Bhadra River and the lower part of Hamkuri River were gradually silted and almost lost their navigability.

7.2.5.5 Mitigation Measures for Shrimp Culture

Technical support and services from the Department of Fishery (DOF) were needed to protect the shrimp culture from damage, but the DOF was reluctant to extend their services to the affected shrimp farmers. For example, the EIA document recommended that a study should be conducted during the implementation of mitigation measures to explore the prospects of the intake of brackish water using small inlets from rivers to facilitate shrimp culture. However, informants said that there was no such study undertaken. The shrimp farmers had to explore the ways on their own in order to protect the shrimp farming from damage and, thereby, their source of income. The informants said that the DOF provided minimum support except imparting training to few community members on shrimp culture.

7.2.5.6 Land Acquisition and Compensation

The EIA report recommended payment of compensation to all affected people whose lands were acquired for the project. The Government acquired 600 acres of land from private land owners. In addition, EIA report proposed Vulnerable Group Development (VGD) program to minimize the sufferings of marginal farmers and to tackle the risk of impoverishment (GOB, 1998). The informants explained that BWDB prepared land acquisition and compensation plan in a timely fashion and placed the plan before the ADB for funding. The ADB provided the funds required for land acquisition and compensation. The payment of compensation was adequate to pay in full to all those who had been affected.

Although the proponent implemented compensation program satisfactorily, the process did not end in time. Land owners had to experience a lengthy and complex bureaucratic process and high transaction costs in order to receive compensation for their land.

7.2.5.7 Implementation of Other Mitigation Measures

From the site visit, direct observation, and a discussion with stakeholders, it appeared that the proponent did not adequately implement other mitigation measures as recommended by the EIA report. These mitigation measures included protection of the aquatic habitat and wildlife from loss, protection of water quality during dredging, plantation of trees, and financial support for farmers' livelihood in the TRM area.

In the case of the plantation program, although there was a plan for the plantation of timber, fruit, and fuel plants in the project area under the guidance of proponents and the Department of Forest, no tangible initiative was noticed by the community people. The action was limited to the homestead areas only rather than other areas, such as road sides. Furthermore, plantation in the homestead areas was undertaken largely by the initiative of the local people themselves. This activity was largely limited to the plantation of rain trees only. The program could be implemented successfully if there were adequate financial and technical support from the proponents and the Department of Forests as indicated by the interviewees.

In summary, the implementation of one mitigation measure, that is, compensation for land acquisition was adequate. The proponent partly

	Proposed mitigation measures	Status of mitigation measures implemented			
		Fully	Partly	Not at all	Unknown
1	Protection of open water fish habitat from loss		■		
2	Management of disposal of dredged materials		■		
3	Protection of water quality		■		
4	Protection of aquatic habitat for plants and wildlife		■		
5	Protection of Bhadra and Hamkura River from possible death			■	
6	Protection of shrimp culture		■		
7	Compensation for land acquisition	■			
8	Financial support for farmers' livelihood in the TRM area		■		
9	Plantation of trees in the project area			■	

Figure 7.1 Status of implemented mitigation measures for KJDRP. (For color version of this figure, the reader is referred to the online version of this chapter.)

implemented other six mitigation measures. Two of the mitigation measures (implementation of TBM option to protect the Bhadra and Hamkura Rivers at their reference condition and the plantation of trees in the project area) were not implemented at all. Figure 7.1 shows the status of the implementation of mitigation measures of KJDRP.

7.2.6 Implementation of Monitoring Program

The proponent (BWDB) introduced third party monitoring system in order to supervise the implementation of mitigation measures and monitor the social and environmental impacts during the project's implementation and operation. The CEGIS was accordingly awarded the contract. However, the CEGIS was appointed to provide monitoring for the first 20 months only, instead of for the full construction period (4 years) of the project. The other 28 months of project implementation was left largely unmonitored. According to the informants, the DOE was supposed to verify the supervision and monitoring records undertaken by the proponent. Initiative from the DOE, however, was not encouraging. Given the limited resources (staff and offices at field level), the DOE infrequently visited the project site.

After the construction was over, the project was operated and managed by the government (BWDB). To monitor the impacts of the project during its operation, the BWDB hired the Institute of Water Modeling (IWM). It had limited expertise of monitoring environmental and social impacts since its service was mainly related to river water management and modeling, and monitoring of engineering aspects. There is now no monitoring activity undertaken either by the IWM or by any institution related to the social and environmental impacts of the project.

At present, the monitoring activity of the IWM is limited to hydrological and morphological changes, such as the depth of rivers, sedimentation rate, salinity, and the width of the rivers. A comprehensive monitoring effort, including environmental and ecological parameters, was required, not only on the river's morphological issues but also in the context of total basin. Moreover, IWM does not maintain a continuous monitoring and recording system. For example, IWM did not set up any office in the project area for tracking impacts of the project. Officials of IWM come from the head office located in Dhaka, the capital city of Bangladesh, to perform their monitoring task. So, there was a lack of continuous and consistent monitoring by the IWM.

7.2.7 Community Participation

Although local community members had an opportunity to participate in the identification of the project's impacts and in the design of mitigation measures, they had only limited opportunity to participate during the implementation of mitigation measures for the project. Discussions with the local community revealed that, since the BWDB did not implement the TBM option according to local community's expectations, the community and other stakeholders (NGOs and civil society) were not fully satisfied. As such, the local community was not interested in being involved further with the implementation of mitigation measures. The project proponent convinced a number of community people including local elites to support implementation of mitigation measures, but most of them had "vested interests."

According to the interviewees, the BWDB involved some local elites and one or two local NGOs. The proponent's intention was to show that it had involved community members in the implementation of mitigation measures and monitoring of implementation of mitigation measures and environmental impacts. However, many of those who became involved had vested interests, were not directly affected by the project, and, therefore, had no credibility to represent the voices of genuinely affected people.

Furthermore, interviewees stated that, during the study of the EIA and the planning of the project, the donor agencies cordially supported community involvement and insisted the proponent involve the local community. However, there was no such support or pressure on the proponent from donor agencies to involve community people for the implementation of mitigation measures. The reason for such reluctance from the donor agencies, however, was unknown.

It was found that the attitude of the project proponent considerably affected community participation during the implementation of mitigation measures. For example, the proponent did not implement the TRM option as per the design and expectation of the local community. The proponent was skeptical about the value of knowledge and experience of the local community. Thus, the proponent was not willing to accept this proposed environmentally sustainable concept.

In the case of KJDRP, although donors had much interest during the EIA study and the design of the project, interest and supervision of equal magnitude were lacking from the agencies during the implementation of mitigation measures.

7.3 CASE STUDY II: JAMUNA MULTIPURPOSE BRIDGE PROJECT

7.3.1 Project Description

Jamuna Bridge, later renamed as Bangabandhu Bridge, built under the Jamuna Multipurpose Bridge Project (JMBP), is the world's 11th longest (4.8 km) bridge. It crosses over the River Jamuna, one of the major rivers of the world. The river has its origin from Assam in India and flows over Bangladesh to the south, physically dividing Bangladesh into east and west parts. The Jamuna Bridge connects the eastern part with the northwestern quadrant of Bangladesh. The location of the bridge is shown in Map 7.2.

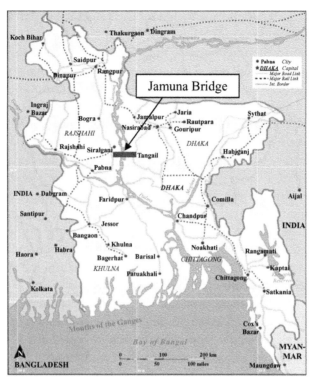

Map 7.2 Location of Jamuna Multipurpose Bridge. (For color version of this figure, the reader is referred to the online version of this chapter.) *Source: Wikimedia Foundation, 2010.*

The Ministry of Communication implemented this large project with the financial and technical support of World Bank and Asian Development Bank during the period 1993–1998. The site is a densely populated rural area. The project area is plain agricultural land and the soil is very fertile. Most people (80%) rely on agriculture for their livelihoods in Bangladesh (BBS, 2005), and this is also true for the project area. Fishing is another important source of income for local people in the project area. Although the site does not have any kind of rich vegetation except homestead forests (a combination of timber, fruit, and fuel trees), the site is rich in many species of terrestrial and aquatic wildlife such as birds, mammals, and reptiles. Importantly, Jamuna River in the project site area is a breeding ground of tortoises (*Batagur baska*) and gharial (*Gavialis gangeticus*) that are listed as threatened species according to the Red Book of International Union of Conservation Nature (IUCN) (IUCN, 1991).

7.3.2 Environmental and Social Impact Assessment of JMBP

The proponent undertook an Environmental Impact Assessment (EIA) study in 1989 as a part of the Project Feasibility Study (JMBA, 1989). Although there was no formal EIA legislation in Bangladesh till 1995, the EIA guidelines for Water Resource Sector was in place in addition to environmental laws (such as Environmental Pollution Control ordinance 1977, the protection and regulations for conservation of fish 1985, Wildlife Conservation Act of 1973, and World Bank's Operative Directives of OD 4.01 and 4.30) for an EIA study of the project (JMBA, 1989).

The EIA study expressed concerns about the hydrological and morphological change of the Jamuna River that could affect the biophysical and social environment in the project area. The EIA team warned particularly that the closure of the northern intake of the Dhaleswari River would have serious impacts on the surrounding natural resources and the society. The EIA report predicted that the volume of water of the river would be significantly reduced due to the closure of northern intake. The reduction of water would adversely affect agriculture, navigation, aquatic habitat, and fish production. Furthermore, the construction of the bridge required a large amount of land to be acquired from the people living in the project site. A significant amount of land was also reclaimed from the Jamuna River at both ends of the bridge. The EIA team assessed that acquisition and reclamation of land in the project area would cause permanent loss of habitat for wildlife. The major significant physical, ecological, and socioeconomic effects are described in Table 7.3.

7.3.3 Proposed Mitigation Measures

To address the impacts identified by the EIA study in JMBA (1989), the proponent devised a rapid Environmental Management Action Plan (EMAP) in 1993 and a detailed action plan of mitigation measures in 1995 (JMBA, 1993, 1995) although the construction of the bridge began early in 1994, including land acquisition and development and reclamation. Furthermore, the proponent prepared a Resettlement Action Plan (RAP) in 1994 (JMBA, 1994). Table 7.4 presents proposed mitigation measures designed against each of the major significant impacts.

7.3.4 Status of Implemented Mitigation Measures

7.3.4.1 Protection of Aquatic Biodiversity

To protect the aquatic habitat from the damage caused by the closure of the northern intake of the Dhaleswari River, the EIA report proposed to create

Table 7.3 Major social and environmental impacts of JMBP Compiled by the authors from JMBA (1989)

	Important environmental issues	Potential significant impacts
1.	Loss of agricultural production and aquatic habitat	a. A total of 7000 ha of land was acquired for the project to construct the bridge and resettle the displaced people. This will cause temporary and permanent loss of agricultural production during and after construction. b. The closure of northern intake of the Dhaleswari River will affect agricultural production in the area due to the reduction of flood water levels. The rate of the recharge of ground water may become lower and availability of water for irrigation during the dry season may be scarce.
2.	Loss of fish production	a. The closure of the northern intake of Dhaleswari River will exert a major impact on fisheries as the water flow and amount of water will be reduced in the river by about 50%. The loss of fish production (492 tons/year) will be reduced by about 25%. b. Around 5650 fisher folks will be directly affected of which about 1400 are full time professional fisher folk.
3.	Loss of wildlife	a. The construction of Jamuna Multipurpose Bridge will intensify human activities on about 7000 ha of acquired land. As a result, the habitat of birds and other wildlife will be permanently lost. The bridge will cause change in the feeding and breeding ground of wildlife. b. Permanent loss of vegetation (homestead forests and bushes).
4.	Disruption of boat navigation	a. Dhaleswari River is a major distributor channel of the Jamuna River which conveys about 15% of flow from the Jamuna toward the Meghna River. Due to the closure of the northern intake, the boats have to detour 60 km when using the southern intake channel and this will cause additional transport cost. A total of 265 families who rely on boating for their livelihoods will be directly affected.
5.	Permanent displacement of affected people	a. About 2166 households will be affected by projects. The people will lose their houses, agricultural land, shops, and other sources of income.
6.	Disruption of water quality in the river during construction	a. Water quality of Jamuna River in the project site will be disrupted due to heavy and long time dredging and other construction activities (piling, river training activity).

Table 7.4 Recommended mitigation measures for JMBP

Proposed major mitigation measures	Mitigation measures
1. Protection of aquatic biodiversity and disruption of navigation	a. Provision of an alternative intake (channel) and enlargement of the southern intake (channel) to offset the loss due to the closure of northern intake of the Dhaleswari River.
2. Increase of agricultural production to compensate for the loss of agricultural production	a. Training to the affected farmers due the closure of Dhaleswari River. b. Provision of credit facilities to farmers: affected farmers will be assisted with credit facilities.
3. Increase of fish production using ponds to offset the loss of fish production	a. Pond fish culture development using culturable and derelict ponds in impact area for project-affected persons (PAPs) and the construction of new ponds. Fish fry and fingerlings will be supplied. b. Provide alternative employment and skill development training on fish culture to PAPs in the project area.
4. Protection of wildlife	a. Awareness program about the value of wildlife will be undertaken during the construction of the project. b. Plantation at both sides of approach roads to provide shelter for migratory birds and a breeding place for many resident birds. c. Establishment of a permanent sanctuary/protected area in one of the existing islands of the Jamuna River to compensate for the loss of wildlife species.
5. Compensation for the disruption of boat navigation	a. Compensation and credit program for directly affected boatmen during the days they have no work. b. Provision of training to pursue alternative employment for boatmen who wish to take different means of livelihood or to supplement their reduced income.
6. Plantation and social forestation (enhancement and compensation for clearance of homestead forests)	a. Plantation along both sides of approach roads (about 26 km) constructed for the bridge. b. Plantation of wood tress in the resettlement sites of 7.5 ha. c. Plantation in the open areas at both ends of the bridge.

Continued

Table 7.4 Recommended mitigation measures for JMBP—cont'd

Proposed major mitigation measures	Mitigation measures
7. Resettlement and compensation for displaced people	a. Compensation for displaced people: cash compensation and displaced people will be resettled in two planned towns to be developed both east and west sides of the Jamuna River and near the ends of the bridge. b. Displaced people (2166 households) will be provided with income generation training to supplement their livelihoods they would lose due to the project.
8. Maintenance of river water quality during construction	a. Dredging will be undertaken based on the seasonal variation.

Compiled by the authors from JMBA (1995).

an alternative channel and enlargement of the southern intake of the Dhaleswari River (JMBA, 1995). However, the proponent did not create an alternative channel and enlarged the southern intake. After the closure of the northern intake, a new off-take opened naturally nearby after 3 years from the start of the project construction works. The proponent assumed that the flow of water through the newly developed channel would offset the reduced flow of water due to the closure of the northern intake.

The Surface Water Modelling Centre (SWMC) of Bangladesh using the North Central Model (widely known as Mike II) analyzed the impacts of the new channel in 2000 and concluded that the channel could send only one-third of the total water flows that had been reduced due to the closure of the northern intake (Imteaz and Hassan, 2001). This shows that the creation of a new channel and the enlargement of the southern intake to protect the aquatic environment from loss were necessary. At present, it is observed that about 60 km of upper Dhaleswari is apparently dead (Photo 7.5). This means that the habitat for fish and other aquatic species is reduced. The loss of available surface water for irrigation is also obvious. The proponent's attention was largely confined to the technical and engineering activities, such as river training and the construction of the bridge. The protection from environmental damage did not get much attention.

Photo 7.5 The fate of upper Dhaleswari (northern intake) in dry season. (For color version of this figure, the reader is referred to the online version of this chapter.) *Source: Photo by Kabir from site visit.*

7.3.4.2 Increase Fish Production to Offset the Loss of Fish Stock

To offset the loss of fish production, the EIA report proposed reform of the existing household ponds in the project areas and to excavate new ponds in the *khas* (government owned) land available in the project area. A training program on how to use new technologies in fish culture was proposed for the affected fishermen. The EIA report also proposed that four hatchery industries would be set up in the project area to meet the supply of fish fingerlings and fish fry for fish culture in the ponds. Moreover, the EIA report proposed for the creation of a revolving fund to run this program sustainably. An NGO called *Grameen Mathsys Foundation* was awarded the contract for implementing these mitigation actions.

However, information from the informants and observation reveal that the proponent partly implemented the mitigation plan. It was observed that the ponds created for fish production were not suitable for fish culture. Most of the ponds dried up and were not reexcavated (Photo 7.6) leading to the abandonment of the ponds. Therefore, the actual production of fish was lower than the target and could not offset the loss of fish production caused by the construction of the bridge.

The plan was suitable in the local context as fish culture in the ponds has been popular and profitable in Bangladesh since the 1980s in order to meet the growing demand of fish consumption. Despite this, the mitigation actions remained unimplemented according to the expected scale and, consequently, anticipated outcomes were not achieved. Some unforeseen difficulties and challenges hindered the implementation of the task. Effective coordination and cooperation between the implementing NGO and the project proponent was required to solve the unforeseen difficulties

Photo 7.6 Excavated ponds (as a mitigation measure) are dried up and abandoned. N.B. Shows the ponds excavated in the *khas* (government owned) land in the project area have dried up and affected community has lost their interest in being involved in fish culture. Currently, most of these ponds are abandoned. (For color version of this figure, the reader is referred to the online version of this chapter.) *Source: Photo by Kabir from site visit.*

and challenges related to the implementation of the task. However, such a coordination mechanism was nonexistent according to most of the informants.

Another problem was related to the timely supply of basic inputs for fish production related to eggs, fish feeds, fish fingerlings, and fish fries. Although the EIA report recommended establishing four hatchery industries to supply fish fingerlings, not a single industry was established by 2000. Before the construction of the project, Jamuna River was the source of fingerlings and fish fries. The construction of the project had a negative effect on the availability of fish fingerlings and fish fries in the locality. Therefore, setting up of hatcheries was an immediate need to meet the loss.

7.3.4.3 Protection of Wildlife from Loss

The proponent undertook a comprehensive inventory for wildlife in the project area before designing the mitigation measures. The study identified that 193 species[5] of the wildlife were available in the project area. Among them, 10 species were listed under the categories of endangered and threatened species by the office of IUCN, Bangladesh (IUCN, 1991). Moreover, the project site was a breeding ground of *gharial* (*G. gangeticus*)

---

[5] There were 9 species of mammals, 169 species of birds, 9 species of reptiles, 6 species of amphibians. Out of 169 species of birds, 50 species were winter migrants, 119 were resident species that were recorded in the Jamuna and Dhaleswari Rivers.

Photo 7.7 *Gharial (Gavialis gangeticus)*: one of the threatened species in Bangladesh. (For color version of this figure, the reader is referred to the online version of this chapter.) *Source: The Daily Star, January 2010.*

and tortoise. These species were also under the category of threatened species (Photo 7.7) according to the IUCN Red Book (IUCN, 1991).

The proponent (JMBA) planned to undertake three mitigation actions in the project area to protect the wildlife from loss. These were as follows: (1) establishment of sanctuaries in the *char* land (land that is raised as the result of accumulation of silt in rivers) near the project site; (2) plantation of trees in the land acquired for the project at both sides of approach roads as well as at both ends of the bridge; and (3) a program to increase awareness about the value of wildlife among the local community and construction workers (JMBA, 1995).

Site visits and information gathered from the relevant sources revealed that the proponent did not create any sanctuary for the potentially affected wildlife species.

Apart from the mitigation measures stated earlier, the EIA report proposed for a mitigation action, particularly for the threatened species (for example, *gharial* and tortoise).

Furthermore, one of the mitigation actions to protect the wildlife was to make the local people and construction workers aware of the value of the wildlife. However, the proponent did not run such an awareness program. Due to the lack of proper initiative to conduct an awareness program to protect the wildlife, hundreds of birds and reptiles were either killed by the workers and local people or died. This came about because of the loss of shelter and food during the development of the project site and construction of the bridge. A museum established on the project site exhibits a sample of this irreversible damage (Photo 7.8). One shortcoming was that the mitigation measures were designed at the time when the land was under the

Photo 7.8 A partial representation of the loss of wildlife during the construction of the project. (For color version of this figure, the reader is referred to the online version of this chapter.) *Source: Photo Kabir from site visit.*

process of acquisition and reclamation. It was necessary to design the mitigation measures before these activities began.

7.3.4.4 Plantation of Trees in the Project Area

Tree plantation along both sides of the approach roads as well as in the project area was one of the important tasks in offsetting the loss of vegetation and homestead forests and the habitat of wildlife. One NGO, called Grameen Bank (GB), was awarded the plantation task for implementation. Project site visits revealed that two-thirds of the total land acquired for the project was left open at both ends of the bridge (Photo 7.9).

7.3.4.5 Resettlement and Restoration of Livelihood for Displaced People

To mitigate the social and economic impact on the displaced people, EIA report recommended compensation measures. Accordingly, the proponent implemented cash compensation and resettlement programs for displaced people in two resettlement areas at both ends of the bridge. More than 2000 households were resettled in two areas at both ends of the bridge. The rest of the people who were not willing to be resettled in these two areas were paid in cash for the loss of their agricultural land, houses, shops, tea stalls, and other sources of livelihoods.

Evaluation report (JMBA, 2001), interview data, and direct observation by the researchers indicate that the implementation of the Resettlement Action Plan (RAP) was largely satisfactory and, hence, the successful outcomes (Photo 7.10). Discussion with people resettled in the project area and direct observation support this claim.

Photo 7.9 Land acquired for the project is left open and underutilized. Source: Photo by Kabir from site visit. (For color version of this figure, the reader is referred to the online version of this chapter.)

Photo 7.10 Resettlement areas for local community displaced by the project. (For color version of this figure, the reader is referred to the online version of this chapter.) *Source: Photo by Kabir from project site visit.*

Many of the displaced people lost their sources of livelihood (such as agricultural land or shops) during the construction and implementation of the project. The proponent recruited them in different construction jobs depending on their skill levels. However, some of the affected people suffered from impoverishment as they had lost their jobs immediately after the completion of the construction of the project. During their employment on the construction of the bridge, however, a good number of workers gained technical experience and became expert on some particular trades, such as dredging and welding. Many of these experienced and skilled people could even manage to get jobs overseas (for example, in the Middle East) after the construction of the bridge.

The presence of donor agencies and their guidance in resettlement were helpful in making the resettlement measure successful, except for the delay in payments in some cases. The timely implementation of compensation was delayed because of the deficiency in land tenure system, in addition to the complex payment system. The selection of the Bangladesh Rural Advancement Committee (BRAC), the biggest NGO in Bangladesh to implement the task, was found to be perfect. The NGO was experienced and committed to implementing the task properly and on time.

7.3.4.6 Implementation of Other Major Mitigation Measures

The EIA report recommended for three other mitigation measures, namely, the measure relating to the increase of agricultural production, the compensation measure for the loss of income for the disruption of boat navigation, and the maintenance of water quality during construction. It was proposed that training on agricultural crops suitable for the Dhaleswari River flood plain would be provided to the affected farmers. Also, credit facilities and supply of seeds for the affected farmers would be arranged. An NGO called *Dishari* was contracted by the proponent to implement the task, in collaboration with the Agricultural Extension Department. According to the informants, few farmers were trained. Moreover, the farmers were not adequately compensated for the disruption and loss of agricultural production during the 4-year construction period of the bridge.

In order to compensate for the disruption of boat navigation, the EIA report proposed three actions: (1) cash compensation for the boatmen for the days they have no work; (2) the provision of training to pursue alternative employment for boatmen who wish to take up a different means of livelihood; and (3) a credit program. However, the JMBA did not implement these actions on the grounds.

The above findings indicate that, out of eight recommended mitigation measures, the proponent completely implemented resettlement program only. Four mitigation measures were partly implemented relating to fish production, the protection of wildlife, tree plantation, and water quality. The proponent did not implement at all the three mitigation measures relating to the protection of aquatic biodiversity, navigation disruption, and the enhancement of agricultural production. Figure 7.2 shows the status of mitigation measures implemented by the proponent of Jamuna Multipurpose Bridge project.

In general, the mitigation measures recommended by the EIA report were inadequately implemented in the case of JMBP. Despite the availability of funds supplied by the donors, mitigation measures, including the

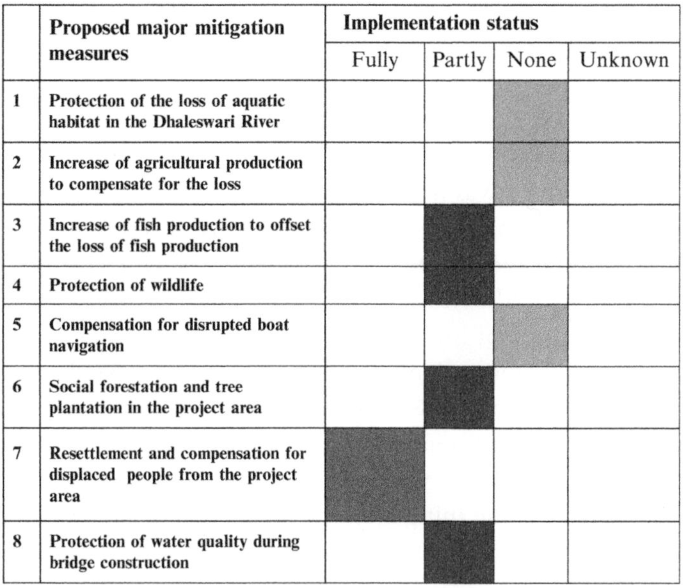

	Proposed major mitigation measures	Implementation status			
		Fully	Partly	None	Unknown
1	Protection of the loss of aquatic habitat in the Dhaleswari River			▨	
2	Increase of agricultural production to compensate for the loss			▨	
3	Increase of fish production to offset the loss of fish production		■		
4	Protection of wildlife		■		
5	Compensation for disrupted boat navigation			▨	
6	Social forestation and tree plantation in the project area		■		
7	Resettlement and compensation for displaced people from the project area	■			
8	Protection of water quality during bridge construction		■		

Figure 7.2 Status of mitigation measures implemented for JMBP. (For color version of this figure, the reader is referred to the online version of this chapter.)

protection of wildlife, production of fish and plantation, and the conservation of aquatic habitat in Dhaleswari were not implemented successfully.

7.3.5 Implementation of Monitoring Program

To implement the mitigation measures successfully, the EIA report proposed for a monitoring program. Accordingly, the proponent established an Environmental Unit (EU) in the project area for the implementation of mitigation measures and monitoring the progress of implementation activities. However, the proponent did not implement monitoring program as to its estimated full coverage and time. The monitoring program was limited only to the river morphology, such as the monitoring of sedimentation, flow of water, and the impact of water flow on the river training embankment around the bridge site. There was no systematic and/or comprehensive monitoring and recording of the mitigation measures relating to social and environmental impacts.

7.3.6 Community Participation

From the beginning of the project planning, members of the local community did not have much awareness about the environmental impacts of the project. This is because the project proponent did not take any noticeable initiative to include the community members during the planning of the project. In fact, there was a lack of initiative by the proponent to include community members during the identification of impacts, designing of mitigation measures, and their implementation. The lack of the proponent's initiative meant that there was no community participation in any form.

Some of the key informants also mentioned that this was the first EIA in the Road Communication Sector in Bangladesh for such a large project. So, the project proponent (the Ministry of Communication) did not have much experience in the study of EIA and in community participation. A number of NGOs, such as Grameen Bank, BRAC, and Grameen Mathsya Foundation, were involved in the implementation of mitigation measures. Usually, NGOs in Bangladesh often represent the voice of rural community people. The above-mentioned NGOs were hired and contracted by the project proponents to implement the mitigation measures. It was unlikely, therefore, that these NGOs, in the role of contractors, would represent the voice of the affected community.

7.4 CASE STUDY III: MEGHNAGHAT POWER PLANT PROJECT

7.4.1 Project Description

The project involved the construction and operation of a gas-fired, combined cycle power plant in Southeast Dhaka, on the northern bank of Meghna River in Meghnaghat, Bangladesh. The site was selected for the construction of three power plants, phase by phase, to generate 2000 MW power. A 450 MW Combined Cycle Gas Turbine (CCGT) power station (Photo 7.11) was constructed in 2002 in phase-I by AES Meghnaghat Limited. The project was constructed at a cost of $295 million and jointly funded by the ADB, WB, and Bangladesh Government.

The site of the project is under the administrative jurisdiction of Sonargaon Upa-Zilla (an Upa-Zilla is the lowest tier of government administrative hierarchy in Bangladesh) of Narayanganj District on 230.77 acres of land (Map 7.3). The project area is surrounded by low-lying fields of paddy and vegetable plots intermixed with clusters of houses. The land was used primarily for paddy cultivation during the dry season and as a fishery during the wet season. A wetland as a part of the low land is situated on the northern side of the project. The wetland is rich in indigenous fish species and is a habitat for birds and other wildlife species. As a predominantly very fertile

Photo 7.11 Meghnaghat Power Plant (back view). (For color version of this figure, the reader is referred to the online version of this chapter.) *Source: Photo by Kabir from site visit.*

area, the project region has seasonal mixed crop vegetation and homestead-based agroforestry.

So as to meet the existing and growing demand of electricity, the Bangladesh government had invited private sectors to invest in electricity generation. The MPPP is such a private sector initiative. The project was completed by 2002 and went into operation in 2003 with 22 years project life (AES International, 2001). The project is currently owned and operated by Pendekar Energy Limited, a Malaysian company.

7.4.2 Environmental and Social Impacts of MPPP

The construction activities of the project involved acquisition of land and dredging in the Meghna River for the development of the project site. The EIA study identified that the acquisition of land for the project would cause the displacement of 17 households. Land development would also

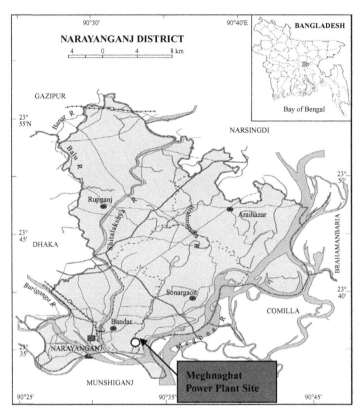

Map 7.3 Location of Meghnaghat Power Plant project site. (For color version of this figure, the reader is referred to the online version of this chapter.) *Source: AES International (2001).*

cause the displacement of wildlife because vegetation in the project area, such as bushes, fuel-wood trees, and fruit trees would be cut down and cleared. The EIA study also identified that the dredging activity would negatively affect water quality, the spawning of fish, and the income of fishermen. Moreover, the construction of the project would affect the nearby wetland, which is rich in indigenous fish, wildlife, and migratory bird species.

The most notable concerns identified by the EIA team were the operation of the project for its 22-year lifetime and the impacts on the surrounding environment and society. The EIA team identified significant environmental and social impacts during the operation of the project relating to water quality, air quality, solid and liquid waste, and noise pollution. Table 7.5 shows the major environmental and social impacts of the project.

Table 7.5 Major environmental and social impacts of Meghnaghat Power Plant project

Important environmental issues	Potential significant impacts
1. Loss of terrestrial and aquatic habitat	a. Loss of habitat, both aquatic and terrestrial, during construction and operation of plant. Trees and vegetation will be lost in the project area due to the MPSA development including construction of access road. b. The project will affect the fish production and habitat of migratory birds in the nearby lowland (*beels*).
2. Noise pollution	During the construction and operation of the plant, noise will occur and disturb the community people living in the adjacent residential area and other sensitive receptors.
3. Air quality	The emission of NO_x and SO_x is inevitable due to the operation of the plant and generation of power. Release of NO_x and SO_x would significantly deteriorate the ambient air quality.
4. Cooling water discharge in the Meghna River water	There would be a significant impact of cooling water on aquatic environment in Meghna River if the wastewater generated through the cooling process is not treated. Also, there could be significant change in population of aquatic organisms and fish production.
5. Impact of solid waste	The plant will use 250 drums each of 45 gallons of lubricating and hydraulic oil each year. The solid waste is generated from the use of lubricating and hydraulic oil mainly in the form of sludge that contain heavy metals such as cadmium, chromium, lead, mercury, and nickel. Therefore, the solid waste will contaminate ground water, surface water, and soil if not treated.
6. Involuntary resettlement	The project site consists of 230.77 acres of land of which 145.515 acres are owned by private landowners. More than 200 household will be affected due to the construction of the project either by the loss of agricultural or homestead land or both.
7. Disruption of water quality during construction	Water quality of the river would deteriorate during dredging in the Meghna River. Fish production and aquatic habitat will be heavily disrupted.

Compiled by the authors from AES International (2001).

7.4.3 Proposed Mitigation Measures

To address the significant impacts outlined in the previous section, a set of mitigation measures were designed and proposed in the EIA report (Table 7.6). Accordingly, the proponent was committed to implement the mitigation measures in order to address the potential social and environmental impacts.

7.4.4 Status of Implemented Mitigation Measures

7.4.4.1 Protection of Terrestrial and Aquatic Habitat from Loss

The EIA team identified that the project would significantly affect the nearby wetland (*beel*) located to the north of the plant. Therefore, the EIA report recommended actions necessary to protect the terrestrial and aquatic habitat in the wetland and adjacent area, for example, releasing fish fingerlings and compensating affected fishermen. A program, including the creation of sanctuary in the northwestern side of the wetland, was proposed to compensate for the loss of habitat and to protect the bird species from disturbance.

Interviews with key informants and observation reveal that, before the project was undertaken, the wetland was in a good condition. One issue was that water from the Meghna River could enter easily into the wetland during the monsoon and high tide time during the dry season. Fish used to come to this wetland from the adjacent Meghna River for breeding and spawning. The wetland was a source of income from fishing for many local people. There were fish of large size available in the wetland, such as *Ruhu*, *Katla*, *Boal*, and *Aire*. The project is located between the river and the wetland (Figure 7.3). As a result, water cannot flow from the river into the wetland (*beel*) easily.

7.4.4.2 Control of Noise Pollution

The proponent has taken necessary actions to protect the neighboring community from noise pollution. It was observed that proponent built a mud wall between the project site and the residential area of the neighboring community. Furthermore, the proponent planted trees to protect the neighbors from noise pollution by the power plant (Photo 7.12). It was found from the site visit that the noise levels did not cross the limit set by the DOE and the World Bank.

Table 7.6 Recommended mitigation measures for MPPP

Proposed major mitigation measures	Mitigation actions
1. Protection of natural habitats from loss	Creation of green belt areas in the project site, including *beels*, to create a habitat for terrestrial and riparian wildlife species and improve aesthetics.
2. Noise pollution control	a. Implement noise management measures. Follow WB, ADB, and Bangladesh Government guidelines. b. Muddy walls will be built around the site to protect neighboring community from noise pollution.
3. Air quality management	a. Use dry-low NO_x burner for the gas turbines. Monitoring to optimize operating load conditions to maximize efficiency and thus minimize emissions. Follow WB and Bangladesh emissions guidelines. b. Use of natural gas and therefore no emission of SO_2. c. Use of 60 m high stack to optimize dispersion of exhaust gases.
4. Management of hot water discharge	a. It was planned that the wastewater, after the proper treatment, will be discharged into the Meghna River. Temperature of effluent discharge meets Bangladesh standards (10 °C) to control the effect on aquatic biota in the Meghna River. b. An effluent disposal system will be installed to effectively treat and discharge of water.
5. Management of solid and liquid waste	a. The sludge generated from storm water system will be collected from the tank and will be analyzed. b. Toxicity of other solid waste generated from lubricants and hydraulic oil will be systematically collected and analyzed by the proponent. c. Then the waste will be delivered to the licensed contractors who have facilities to treat the waste and to recycle the treated waste for other purpose.
6. Plantation of trees	Trees will be planted in the open space of the project area to compensate for the loss of vegetation and habitat for birds and other wildlife.
7. Implementation of involuntary resettlement	Assistance to households for loss of houses or land by way of cash compensation. Moreover, project-affected people will be recruited during the construction and operation of the project to offset their income loss.
8. Management of water quality	In the winter season and at low tide, dredging activity should be suspended for some time. Water quality will be closely monitored during dredging.

Compiled by authors from AES International (2001).

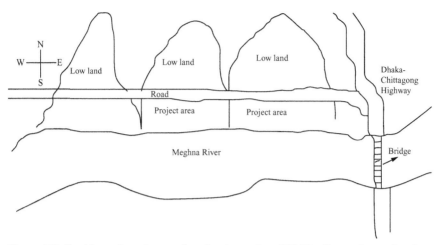

Figure 7.3 The disruption of water flow by the project. N.B. The figure shows that free flow of water from Meghna River to low land (wetland) is being interrupted because of construction of road. *Source: Drawn by Kabir based on site visit.*

Photo 7.12 Constructed mud walls and planted trees. (For color version of this figure, the reader is referred to the online version of this chapter.) *Source: Photos by Kabir from site visit.*

7.4.4.3 Air Quality Management

From observations and information gathered from key informants, it is evidenced that the proponent installed modern technology to manage air quality. The emission of SO_x and NO_x pollutants was a matter of concern. To control the emission of NO_x, the proponent installed dry-low NO_x gas turbines. Two stacks of 60 m in height each were also built to emit the NO_x into the air (Photo 7.11) as recommended.

7.4.4.4 Management of Hot Water Discharged into the Meghna River

According to the recommendation by the EIA report, the proponent planned to discharge the waste hot water into the Meghna River after its proper treatment. The temperature of the effluent discharge would meet Bangladesh standards (10 °C) in order to control the effect on aquatic biota in the Meghna River. With this in mind, the proponent installed an effluent disposal system to treat the hot wastewater generated from cooling process. The officials of the project office have mentioned that they used to release hot water into the Meghna River after the hot wastewater is properly treated. However, it was not possible for the researcher to check the quality (such as toxicity and temperature) of the cooling water released into the river.

The local community in the project area said that they had little information about whether the proponent properly treated wastewater before it was discharged into the river. The project proponent has always claimed that it discharges toxic hot water after proper treatment. However, some past incidents (no fish in the area, buffalo dying after drinking water, a man nearly died from the hot and toxic water after his boat had sunk in the vicinity) demonstrate the futility of the proponent's claims.

7.4.4.5 Management of Solid and Liquid Waste

The proponent installed an Effluent Treatment Plant (ETP), as observed by the researchers, to manage the solid and liquid waste generated from the plant. The proponent also established an Environmental Health and Safety Unit. This included a laboratory for conducting environmental management activities related to operating the ETP, and collecting, testing, and disposing of solid and liquid waste. It was noted by way of observation that the project proponent had put due efforts into managing contaminated waste released by the power plant. The treated waste is collected and disposed of by the contractors as approved by the DOE.

Despite the satisfactory arrangement for the management of solid and liquid waste, the informants, however, expressed their concerns that the waste management activities undertaken by the proponent should be closely and regularly monitored and verified by the DOE. In the absence of the DOE's regular monitoring, the project proponent may not operate ETP at all and release toxic waste into the river water. This is a common phenomenon in Bangladesh where industries often tend to release toxic waste without treatment. An ETP is also a costly operation. Therefore, particularly during the rainy or flood seasons, many industrial proponents tend to discharge the effluents directly into water without any treatment. The researchers observed that this concern was compounded among the community people surrounding the MPPP. The community people have no easy access to the project site to check the environmental management activity of the project proponent.

7.4.4.6 Plantation of Trees in the Project Area

The EIA report recommended plantation of trees in the project area to compensate for the loss of vegetation that occurred during the construction of the project. However, during the site visit, the researchers observed that the plantation of trees covered only a little area of the project site. The vast amount of land of the project site has been left unused (Photo 7.13). According to the informants, the plantation of wood, fuel, and fruit trees in this open land could be a source of food and habitat for wildlife and fuel-wood for the affected community. According to the officials of the project, since the site was developed by the government, it was the responsibility of the government to green this open area.

7.4.4.7 Involuntary Resettlement and Compensation for Land Acquisition

The government office acquired a total of 237.22 acres of land for the project. The EIA report recommended that the affected people be paid in cash for the loss of their land. It also recommended that the local people should be recruited according to their skills during the implementation and operation of the project. Furthermore, the affected people should be given training on self-employment activities.

Photo 7.13 Acquired but unused land in the project area. (For color version of this figure, the reader is referred to the online version of this chapter.) *Source: Photos by Kabir from site visit.*

The compensation for the loss of land was paid on time to all affected individuals. However, the recipients, i.e., the land owners, were not happy since the payment was lower than the actual market price. The price of the land acquired was much lower than the actual market price at that time. It was not possible for the affected people to buy the same amount of land adjacent to the project area with the money they were paid.

The key informants said that the proponent imparted training to affected people on self-income–generating activities. The proponent also offered jobs to the affected people during the construction of the project. The proponent employed about 100 people from the local community as temporary semiskilled and unskilled laborers.

After the completion of the project, however, many of the temporarily recruited project-affected people became unemployed again. The proponent recruited many semiskilled and unskilled people to operate the plant. However, discussion with the informants reveals that the proponent did not recruit anyone from the affected community.

7.4.4.8 Management of River Water Quality

During the construction of the plant, the proponent took appropriate measures to protect the water quality of the Meghna River. So, there was a possibility of surrounding river water being polluted because of dredging. There was close monitoring to observe the water quality, as the project officials said. More importantly, during the winter season (dry season) and low tide, dredging activity was suspended for some time. The Bangladesh Inland Water Authority regularly supervised this activity. The River Meghna is a big river, so the quality of water in the river did not deteriorate during the construction of the project.

In summary, the findings analyzed above show that among the eight mitigation measures, the proponent implemented four mitigation measures relating to noise pollution, air quality, solid waste, and management of water quality during construction fully. The other four mitigation measures relating to involuntary resettlement and compensation, the plantation of trees, the protection of river water quality from dredging, and the aquatic and terrestrial habitat were implemented partly. Figure 7.4 shows the status of implementation of mitigation measures.

7.4.5 Implementation of Monitoring Program

According to the project officials, the proponent conducts monitoring as recommended by the EIA report. The proponent submits a monitoring

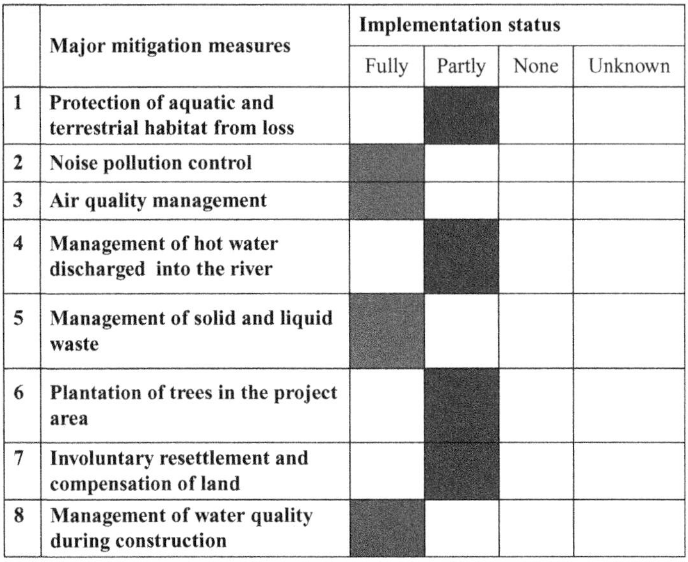

Major mitigation measures	Implementation status			
	Fully	Partly	None	Unknown
1 Protection of aquatic and terrestrial habitat from loss		■		
2 Noise pollution control	■			
3 Air quality management	■			
4 Management of hot water discharged into the river		■		
5 Management of solid and liquid waste	■			
6 Plantation of trees in the project area			■	
7 Involuntary resettlement and compensation of land			■	
8 Management of water quality during construction	■			

Figure 7.4 Status of mitigation measures implemented for MPP. (For color version of this figure, the reader is referred to the online version of this chapter.)

report to the DOE every month. The proponent also sends an environmental performance report to the ADB and the DOE annually. The proponent has also established an in-house laboratory to test the quality of cooling water and other solid and liquid waste. The proponent hired a third party, an organization called Resource Control Centre (RCC) to monitor the aquatic condition of the Meghna River. Every 2 years, the RCC surveys quality of water in the Meghna River around the project site. Furthermore, air quality is monitored twice a year by air quality experts.

Although the proponent has established an in-house monitoring program, concerns from key informants (e.g., local community) suggest that this cannot guarantee that the proponent is adequately performing monitoring program. This is because proponents often may manipulate their monitoring data if the activity is not frequently checked by the DOE. In practice, industrial proponents often send samples of water quality, for example, to the DOE that are often found to have been manipulated and misleading. Therefore, the DOE staff should regularly visit the project site and verify the proponent's mitigation and management activities on the ground instead of performing a "desktop exercise" as they often do.

Another technique of monitoring applied by the proponent of MPPP was the systematic recording of complaints submitted by the local community about the environmental pollution. However, there was no such systematic recording system either in the project office or elsewhere, such as in local government offices. Additionally, the environmental court or environmental office is not available in district level, let alone Upa–Zilla (subdistrict) level. So, there were hardly any complains made by the local people. Also, in general, local people in Bangladesh do not make any complaints unless they are directly affected by any project. Often, they do not make complaints about the pollution for fear of harassment by the influential project proponents.

7.4.6 Community Participation

During the EIA, the community was involved in the form of consultation through formal and informal meetings. A socioeconomic survey was conducted among the households of three local villages (Islampur, Ganga Nagar, and Dhadghata) to collect socioeconomic data. This survey gave the local community an opportunity to learn about the project and its activities that may affect the surrounding environment and society. The draft EIA report was also placed in the Local Council (lowest tier of local government structure) where those people affected by the project and other stakeholders had the opportunity to make comments on the potential impacts of the project.

While the proponent involved the concerned community during the EIA study and the development of the EIA report, the involvement of community in environmental management during the operation of the project was limited. The proponent formed a committee consisting of people affected by the project called Project Affected Persons' (PAPs) committee. Therefore, local people had an opportunity to meet and negotiate with the project proponent by means of this committee. As the representative body of affected local people, this PAPs committee sits with the proponent every 4 months. The proponent informs the members of PAPs committee of mitigation and management of impacts of the plant's operation.

However, information from the site visit and discussion with the community people reveal that members of PAPs committee take part in the meeting merely as listeners without having any scope to put their concerns or verify the proponent's claims.

7.5 EIA PRACTICE AT POST-EIS STAGE: OVERALL ENVIRONMENTAL MANAGEMENT PERFORMANCE OF THE THREE PROJECTS

7.5.1 Implementation of Mitigation Measures of Three Projects

The status of the implementation of the major mitigation measures adopted in the three projects is shown in Figure 7.5. Among the eight major mitigation measures of JMBP, 12% (one) measures were fully implemented, 50% (four) measures were partially implemented, and 38% (three) were not implemented at all. In the case of KJDRP, out of nine mitigation measures, 11% (one) measure was fully implemented, while 67% (six) measures were partly implemented. The remaining 22% (two) measures were not implemented at all. In the case of MPPP, among eight mitigation measures, 50% (four) mitigation measures were fully implemented and 50% (four) mitigation measures were partly implemented. Overall, the findings show that no project achieved its environmental management targets in terms of the implementation of recommended mitigation measures. Due to the inadequate implementation of mitigation measures, as revealed in the case of three projects, the significance and effectiveness of EIA as a flagship of environmental management tool largely remained underappreciated.

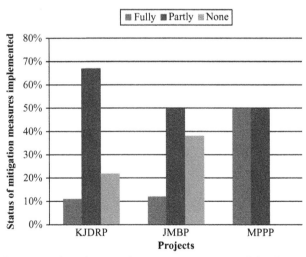

Figure 7.5 The status of implemented mitigation measures of the three projects. (For color version of this figure, the reader is referred to the online version of this chapter.)

7.5.2 The Role of DOE

This study shows that, during the implementation of the mitigation measures, the DOE had exercised inadequate supervision to track the implementation of mitigation measures and management of environmental impacts. Although the DOE is empowered to take strong punitive actions against the proponents for inadequate implementation of mitigation measures, the DOE did not take any stern action. This includes no action in lodging a case against the proponents of KJDRP or JMBP, except serving show cause notices. This is because, according to the informants, there is an unwritten rule in Bangladesh that one government agency cannot sue another government agency. The informants also expressed their concerns that this immunity may lead a public proponent to evade the project approval conditions by failing to adequately implement the mitigation measures.

On the other hand, DOE officials reveal that, given the limited financial and technical resources, such as budget, staff, expertise, and laboratory equipment, the DOE could not adequately supervise the implementation of mitigation measures. However, according to the informants, the lack of sincerity among the DOE officials in regularly visiting the project sites contributed more to the poor supervision than the scarcity of resources. Officials become proactive in supervising mitigation measures only when environmental pollution is published in the print or electronic media.

7.6 CHAPTER SUMMARY

This chapter presents findings on the effective EIA practice at the post-EIS stage of EIA system in Bangladesh. The main finding is that EIA practice at the post-EIS stage of EIA process in Bangladesh is inadequate. Results show that mitigation measures were not adequately implemented in the case of three projects analyzed in this chapter. Furthermore, community participation and monitoring activities required for effective implementation of mitigation measures were found to be inadequate.

REFERENCES

AES International. Environmental impact assessment: Meghnaghat power project, Dhaka: Bangladesh; 2001.
Bangladesh Bureau of Statistics (BBS). Handbook on environmental statistics. Dhaka: Ministry of Planning, Government of Bangladesh; 2005.

Cashmore M, Gwilliam R, Morgan R, Cobb D, Bond A. The interminable issue of effectiveness: substantive purposes, outcomes and research challenges in the advancement of environmental impact assessment theory. Impact Assess Project Appraisal 2004;22 (4):295–310.

Chakraborty TR n.d. Management of Haors, Baors, and Beels in Bangladesh: lessons for Lake Basin management. Viewed 13 March, 2011 http://wldb.or.jp/ILBMTrainingMaterials/resourcesBangladesh.

Government of Bangladesh (GOB) . Khulna-Jessore Drainage Rehabilitation Project, Bangladesh Water Development Board. Ministry of Water Resource Development, Government of Bangladesh; 1998.

Imteaz MA, Hassan KI. Hydraulic impacts of Jamuna Bridge: mitigation option. J Environ Eng 2001;422–428.

International Union of Conservation Nature (IUCN). Red book of threatened and endangered species in Bangladesh. Dhaka, Bangladesh: IUCN Country office; 1991.

Jamuna Multipurpose Bridge Authority (JMBA). Jamuna Bridge Project, Feasibility Report, Phase 2 Study, vol. 1. Main Report, Ministry of Communication, Dhaka; 1989.

Jamuna Multipurpose Bridge Authority (JMBA). Jamuna Bridge Project Environmental Assessment (Compilation of Summaries). Dhaka: Ministry of Communication, Government of Bangladesh; 1993.

Jamuna Multipurpose Bridge Authority (JMBA). Resettlement Action Plan. Dhaka: Ministry of Communication, Government of Bangladesh; 1994.

Jamuna Multipurpose Bridge Authority (JMBA). Environmental Management Action Plan. Dhaka: Ministry of Communication, Government of Bangladesh; 1995.

Jamuna Multipurpose Bridge Authority (JMBA). Resettlement in JMBP: Assessing Process and Outcomes, Qualitative Evaluation of RRAP (Revised Resettlement Action Plan) and Project on EFAP (Erosion and Flood Affected Persons). Dhaka: Government of Bangladesh; 2001.

SMEC International. Khulna-Jessore drainage rehabilitation project, project completion report. Dhaka, Bangladesh; 2002.

The Daily Star online edition. Dhaka, http://thedailystar.net/newDesign/index.php; 2010 [accessed 22 December 2011].

Wikimedia Foundation. http://upload.wikimeadia.org/Wikipedia/commons/b/bd/map; 2010 [accessed 13 March 2011].

Wood C. Environmental impact assessment—a comparative review. London: Prentice Hall; 1995.

CHAPTER 8

Evaluating Environmental and Social Impact Assessment: Framework for Effective EIA System

Contents

8.1 INTRODUCTION

This chapter is a synthesis of findings presented in the preceding five chapters. Section 8.2 discusses the current institutional arrangements of environmental and social impact assessments (EIAs) in Bangladesh and the possible causes of current deficiencies in institutional arrangements. Section 8.3 briefly discusses the quality of environmental impact statements (EISs) and the factors influencing the quality of EISs. Sections 8.4 and 8.5 provide overview of consideration of social issues in EIA and community participation, respectively. The practice of EIAs at the post–EIS stage in terms of the implementation of mitigation measures and monitoring activities are discussed in Section 8.6. Possible reasons for the inadequate implementation of mitigation measures are analyzed.

Evaluating Environmental and Social Impact Assessment in Developing Countries © 2013 Elsevier Inc.
http://dx.doi.org/10.1016/B978-0-12-408129-1.00008-5 All rights reserved.

Section 8.7 focuses on the overall status of the EIA system in Bangladesh. Section 8.8 presents a proposed framework for effective EIA system that has emerged from this study. This chapter concludes with a brief summary of the findings and makes recommendations for improvement in the EIA system.

8.2 INSTITUTIONAL ARRANGEMENTS FOR THE EIA SYSTEM IN BANGLADESH

The findings in Chapter 3 show that significant efforts have been made throughout the past two decades to establish a viable institutional setup for EIA practice in Bangladesh. Old legislation (Environmental Protection Ordinance of 1977) was repealed and new legislation (Environmental Conservation Act of 1995 and Environmental Conservation Rules of 1997) was enacted with a greater scope for environmental management. This new legislation provides the legal foundation for the EIA system in Bangladesh.

As part of the increased emphasis on environmental protection, there has also been an initiative to reorganize the administrative arrangements for the application of EIAs. The pollution control cell created under the Environmental Protection Ordinance in 1977 has now grown into the Department of Environment (DOE). Now there is a separate Ministry of Environment and Forest (MOEF) responsible for making environmental policies and legislations. Within the MOEF, the DOE carries out its responsibilities to implement EIAs. In addition, the establishment of an Environmental Court in 2000 is another initiative of the government to make the EIA system effective. These institutional arrangements represent a key feature of an effective EIA system as proposed by authors, such as Abracosa and Ortolano (1987), Ortolano et al. (1987), and Ortolano (1993).

However, the presence of mere legal provisions and organizations is not enough to have an effective EIA system. The legal provisions need to be clearly articulated with the adequate delineation of EIA requirements. The legislation of EIAs needs to be comprehensive and enforceable in practice. It needs to be comprehensive so that proponents are bound to implement the requirements.

The findings in Chapter 3 show that the provisions of the EIA requirements are not comprehensively and clearly stated in current legislation (ECA of 1995 and ECR of 1997). Also, the review process of EISs and the provision of EIAs for modified or extended projects are not detailed adequately by the legislation. Due to the lack of comprehensive rules for EIA requirements, proponents may avoid some requirements during the implementation of EIAs.

Furthermore, there is a lack of clarity in the EIA requirements legislated by the Environmental Conservation Act of 1995 and Environmental Conservation Rules of 1997. This may create confusion and uncertainty among the proponents in effectively applying EIAs in Bangladesh. The lack of clarity and comprehensiveness in the EIA legislation (ECR) can be attributed to the fact that the EIA is relatively new in Bangladesh. In addition, after the introduction of EIA laws in 1995, there were no comprehensive reviews by the government or academics and there were no major amendments or modifications made to the laws. It is, therefore, not surprising that some deficiencies in terms of comprehensiveness and clarity of provisions exist in current EIA legislations.

While the DOE is legally empowered by the legislation to enforce the EIA, it lacks adequate budget and staff. Moreover, the lack of stable leadership and the absence of a decentralized administration are major weaknesses. As a relatively new organization, the DOE's position in the government bureaucratic hierarchy is not well established as yet. Hence, the DOE has limited influence over politically influential proponents. These weaknesses limit the DOE's enforcement capacity and the scope of the delivery of intended services. Previous studies (for example, Briffett, 1999; Doberstein, 2003) have found that the relatively weak position in the government bureaucratic hierarchy and the lack of adequate resources often limit the environmental agency's control over the proponents.

The reason for the DOE being under-resourced in terms of its budget and manpower can be attributed partly to the current socioeconomic, political, and bureaucratic context of Bangladesh. It is widely recognized that, due to the high standard of living and education in the industrialized nations, there is a great deal of support for environmental protection from the general public, which has resulted from a high degree of awareness of environmental matters. On the other hand, in developing countries, the standard of living and education is still very low and environmental issues are accorded a much lower priority. Therefore, the approach to environmental protection in developing countries is top-down and the success of those administrative arrangements that have inadequate resources for implementing EIAs often lies in convincing policy makers (particularly politicians) of the importance of environmental protection.

The DOE depends on the Ministry of Finance for its budget. There is a conflict of interest between the two bodies. Ideally, as the custodian of environmental management, the DOE advocates for sustainable development where economic growth is encouraged. On the other hand, the Ministry of Finance, as in many other developing countries, tends to rate the attractive figure of economic growth more highly, even though it might be at the cost of the environment. The people of Bangladesh do not have adequate awareness of environmental issues, and, therefore, pressure from the affected people is often not significant. Thus, the current socioeconomic and political context in Bangladesh puts constraints on the capacity of the DOE to implement EIAs effectively.

The weaknesses of institutional arrangements (legal and administrative) identified in this study are, however, not unique to Bangladesh. Previous studies in developing countries (Ahmad and Wood, 2002; Briffett, 1999; Glasson and Salvador, 2000; Lee and George, 2000; Sadler, 1996) have similar findings. This study suggests that an EIA system with comprehensive legislative provisions for EIA application is required in Bangladesh. Some provisions of EIA requirements (for example, provisions regarding the review of EISs, an EIA of the extension of an old project, and the stages of the EIA process) need to be incorporated. Also, some existing provisions of EIA requirements (for example, the validity of ECCs, the provision of site clearances, and the community's right to go directly to court) need to be amended to address ambiguity. A complete form of legislation, with clear provisions for EIA requirements, is necessary in Bangladesh. Many developing countries, for example, the Philippines (ADB, 2007), Thailand (Memon, 2000), Brazil (Glasson and Salvador, 2000), and China (Wang et al., 2003) have amended their EIA legislation such that it contains comprehensive provisions for the requirements and broader scope of EIA applications.

The findings of this study (Chapter 3) also suggest that, in order to make the EIA system effective, the administrative setup in Bangladesh must be sufficiently capable in forcing and guiding project proponents to comply with EIA requirements. On the one hand, this requires the DOE to have adequate resources (budget and staff) and expertise to perform the intended tasks. On the other hand, the DOE also requires a stable leadership and a decentralization of its function at local level (district and subdistrict levels) through the setup of offices. It is a positive sign that, recently, the government has decided to increase the manpower of the DOE from 267 to 735 and it is hoped that the DOE will be able to cover the current shortage of manpower.

8.3 THE QUALITY OF EISs

The quality of EISs reflects the effectiveness of the EIA process in practice. The findings in Chapter 4 show that the quality of EISs in Bangladesh in general can be classified as being merely satisfactory, while a significant number (34%) of them are unsatisfactory. This also suggests that some tasks (criteria) need to be performed well as many have been found to be poorly performed. These findings are broadly in line with previous studies conducted in developing countries and elsewhere (for example, Jalava et al., 2010; Sandham and Pretorius, 2008). For example, Barker and Wood (1999) found 50% of the reviewed EISs to be unsatisfactory and Badr, Cashmore, and Cobb (2004) found 32% of the reviewed EISs to be unsatisfactory.

The quality of an EIS is found to be deficient when important criteria (tasks) in the EIA reports are not well addressed or are omitted. This study shows in particular that the tasks such as baseline data, impact prediction and assessment, the design of mitigation measures, and alternative analysis and monitoring are often poorly performed. These findings on poor performance concur with the findings of previous studies (for example, Fuller, 1999; Glasson et al., 1997; Sadler, 1996; Sandham and Pretorius, 2008). Even in a mature EIA system, the quality of EISs suffers from inadequacies, especially with regard to impact prediction, the determination of significant impacts, and the design of mitigation measures (Fuller, 1999; Sadler, 1996).

The reviewed EISs include lack of adequate, systematic, and accurate baseline data; poor impact prediction; and poor evaluation of significant impacts. This situation indicates that the scientific aspects of EIAs are poorly addressed in Bangladesh. During the review of the quality of EISs, it was observed that where the EISs' baseline data were poorly presented, subsequent actions relating to the prediction of impacts and mitigation measures were also poor. This suggests a positive relation between baseline data, impact prediction, and mitigation measures (Figure 8.1).

The deficiency in the quality of information that is contained in EISs in Bangladesh is attributable to a number of immediate factors. These factors include inadequate time given for EIA study, the commercial interest of consultants and proponents, a lack of EIA experts, defective service procurement for public projects, a lack of funds, weak TORs, and weak EIA teams.

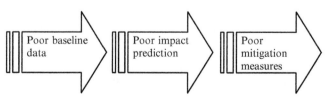

Figure 8.1 The relationship between baseline data, impact prediction, and mitigation measures. *Developed by the authors.*

These findings concur with the studies of Lohani et al. (1997), Modak and Biswas (1999), and Morrison-Saunders et al. (2001). For example, Morrison-Saunders et al. (2001) underscore that time and financial resources are among the main driving forces of EIS quality. Authors such as Modak and Biswas (1999) and Lohani et al. (1997) hold the view that a clear and complete TOR is an important factor in a good quality EIS.

In addition to the above factors, the deficiency in comprehensive legal requirements relating to EIS preparation and a lack of sector-specific and technical guidelines may influence the quality of EISs. As mentioned earlier (Chapter 3), there are only four sectoral EIA guidelines available. There are no technical guidelines for baseline data collection and for prediction and assessment of impacts using suitable techniques. In the absence of adequate sector-specific and technical guidelines, EIA practitioners often face difficulties in determining the right type of data required, the application of appropriate models, and the techniques for predicting and evaluating impacts.

This study suggests that the provisions relating to the code of conduct and accreditation system for qualified EIA consultants are required to improve the quality of EISs in Bangladesh (Momtaz, 2002). At present, EIA consultants are paid by the proponents and, therefore, consultants often tend to prepare the EISs according to the will and objective of the proponents. The proponents' objective is often to prepare the EIS to obtain environmental clearance from the DOE. The regulatory control of consultants through the introduction of a code of conduct is, therefore, necessary to force EIA consultants to prepare EISs with minimum bias and, thus, improve the practice.

8.4 CONSIDERATION OF SOCIAL ISSUES IN IMPACT ASSESSMENT

Like many developing countries, in Bangladesh, SIA is conducted as an integral part of EIA (Chapter 5). The term "social impact assessment" (SIA) has not been included in the EIA legislation. However, the definition of environment provided in legal documents includes social aspects of the environment; thus, EIA also includes assessment of social and economic impacts. The DOE has also clearly identified "Human" as an important environmental component in its guidelines. The guidelines of the donor agencies have outlined the processes of social impact consideration and emphasized the need for social soundness of projects. The quality of SIAs is generally good. However, to further strengthen the concept of social impacts into EIA, there should be clear legal mandate for SIA. The donor agencies should continue to play a supervisory role in the implementation of EIA, SIA, and community participation. There should be a major review of SIA in Bangladesh. SIA is a relatively newcomer in developing countries. Often, it constitutes a chapter in EIA reports. Here, SIA processes and effectiveness have not received much scrutiny to date. The limited studies (see Chapter 6) we have seen in this field for developing countries are in agreement that SIA is an emerging field in developing countries and will need concerted efforts by the governments and the NGOs for it to be more effectively integrated into development planning. For developed countries, on the other hand, SIA is seen to be evolving from a mechanical process of identification of social impacts and development of mitigation measures to a process of social change (Vanclay and Esteves, 2011).

8.5 COMMUNITY PARTICIPATION IN EIA AND SIA

Community participation is not mentioned in the legal documents, but it is well established in the DOE and donor guidelines (Chapter 6). Proponents tend to include local community in the decision-making process and in the development of management and mitigation plans. However, there is no meaningful participation from the affected parties at the post-EIS phase. It appears that the DOE and the donor agencies lack control over what happens after the EIA has been approved and the proponent has proceeded with the project implementation. It is therefore imperative that community participation is given a legal mandate via its inclusion as a separate clause in the amended ECA '95. It is also important that the DOE is strengthened for it to be more effective in the post-EIS phase and that the donor agencies maintain their interests in their funded projects even after the completion of EIA.

8.6 EIA PRACTICE AT THE POST-EIS STAGE: IMPLEMENTATION OF MITIGATION MEASURES

The key features of an effective EIA system include the adequate implementation of mitigation measures and monitoring at the post-EIS stage. However, in the case of the three projects (details in Chapter 7), the findings show that there is a clear gap between the mitigation measures recommended by the EIA reports and the mitigation measures actually implemented by the proponents. In general, mitigation measures have been only partly implemented in all three projects. Also, there have been inadequate community participation and monitoring (Figure 8.2). Since the mitigation measures

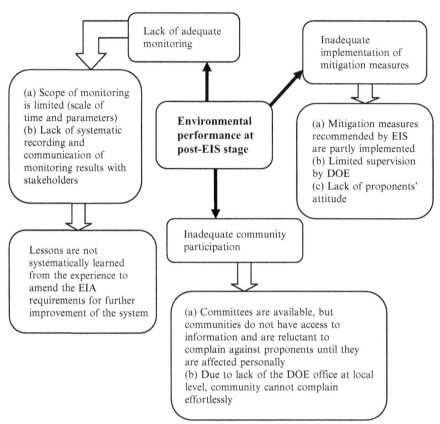

Figure 8.2 Environmental management performance of the three projects at post-EIS stage of EIA process. *Developed by the authors.*

have been only partly implemented, it can be presumed that the targets for environmental protection of the projects were not fully achieved.

A number of similar studies conducted in the past, for example, Morrison-Saunders, Baker and Arts (2003), and Ortolano and May (2004) also have demonstrated that the implementation of mitigation measures and other activities is poorly performed at the post-EIS stage in the EIA process.

Although there has been an inadequate implementation of mitigation measures in the case of the three projects, the performance of the Meghnaghat Power Plant Project (MPPP) is better than the other two projects (KJDRP (Khulna-Jessore Drainage Rehabilitation Project) and Jamuna Multipurpose Bridge Project (JMBP)). The reasons for the better performance may be that the private proponent was careful about its corporate image as an international company. As a part of this, a laboratory was established in the project office and the project office appointed a full-time environmental health and safety officer to look after environmental issues. Also, the proponent sends environmental management reports voluntarily to the DOE and sits with local community people once every 2 months. On the other hand, in the cases of the KJDRP and JMBP, the proponents were government (public) agencies and did not implement the mitigation measures with due attention. The primary reason for this negligence could be the unwritten rule that, as a government agency, the DOE cannot sue another government agency.

In the case of the three projects, findings also reveal that active participation by the community people was not noticeable during the implementation of mitigation measures. In the case of the KJDRP, the community had an active role as well as the support of donor agencies in designing the project. However, the community's role was limited during the implementation of the project as its members were not happy when they saw that the proponent (Bangladesh Water Development Board) was reluctant to implement the project according to their proposed design. In the case of JMBP, the community had very limited active participation during the implementation of mitigation measures (except the resettlement action). In the case of the MPPP, community participation has been in place in the form of a committee. This shows that the proponent is accountable to the neighboring community, but the findings also show that the proponent only informs them about their own activities rather than encouraging active participation.

The findings also reveal that mitigation measures could be implemented more effectively by actively involving community people given their special environmental knowledge. While experience shows that members of the communities may not have a good understanding of biophysical impacts, in some cases, they have a better understanding of the socioeconomic impacts of projects and the ways to mitigate them than the EIA experts (Momtaz, 2006). Therefore, both the proponents and regulators in Bangladesh need to realize the potential strengths of the community members that may be useful in effectively implementing mitigation measures.

The poor implementation of mitigation measures in the case of the three projects can be attributable to a number of factors. In the case of all three projects, these factors include the attitude of the proponents, the lack of coordination between actors involved in the implementation of mitigation measures, limited supervision by the DOE, and limited support from the donor agencies.

In Bangladesh, in general, the proponents perceive that the purpose of an EIA is just to prepare reports and gain an ECC from the DOE. This short-term objective may divert proponents' attention away from the effective implementation and management of mitigation measures. This finding confirms the views of other studies, for example, by Dipper et al. (1998) and Sadler (1988). The lack of attention by the proponents in the post-EIS stage of EIA is a major constraint in the advancement of EIA practice in Bangladesh.

Similarly, donors' support might contribute in many ways to the adequate implementation of mitigation measures. However, there is a lack of support from donor agencies at the post-EIS phase. Like many other counties, proponents prepare Project Completion Reports (PCRs) and send those to the donor agencies. However, these reports reflect only bare detail whether mitigation measures were implemented adequately. The main focus of the reports is merely the completion of physical construction. The results can be improved if the donor agencies evaluate the PCRs, visit the project sites, and discuss the PCRs with the affected communities. However, this is not the case in Bangladesh. In general, the donor agencies tend to think that it is the sole responsibility of the government to oversee the environmental performance of the proponent.

The practice of EIAs in Bangladesh is still largely preoccupied with the preparation of EIA reports. This is illustrated by the poor implementation of mitigation measures along with inadequate monitoring, limited community involvement, minimal supervision from the DOE, the proponents' narrow attitude, and the low level of support from the donor agencies. This means that the post-EIS stage of EIA systems remains a relatively neglected area, despite the fact that the ultimate effectiveness of an EIA system depends on the effective implementation of mitigation measures.

While it is possible for the DOE to control the proponents during the preparation of EISs, it has limited control on proponents at the post-EIS stage during the implementation of mitigation measures.[1] Thus, having a legal authority to enforce the proponents to implement mitigation measures is not enough. Given the socioeconomic and political situation, the DOE needs to play its role as a persuader[2] as well as an enforcer. As a persuader,

---

[1] It is easy for the DOE to control the proponents at EIS preparation stage because the proponents need to go to the DOE to submit an EIS and obtain an ECC. The DOE may reject the poor quality of an EIS and this may delay the environmental clearance. On the other hand, after the ECC is obtained, proponents are not under direct control as in the time of EIS preparation and, usually, the DOE cannot sue against the public proponents if they do not comply with environmental clearance conditions (such as the implementation of mitigation measures).

[2] The persuasion theory is concerned with the shifts in attitude. Persuasion is typically defined as human communication that is designed to influence others by modifying their beliefs or attitude. Persuasion is not accidental or coercive; it is inherently communicational.

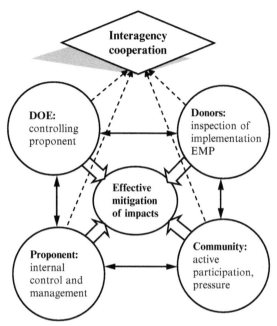

Figure 8.3 Interaction of four key parties in the implementation and management of mitigation measures. *Developed by the authors.*

the DOE may convince the politically powerful project proponents by modifying their beliefs, values, and attitudes toward the benefits of environmental management.

This study emphasizes that an effective coordination mechanism is required in Bangladesh that allows the proponents, regulators, public, and donors to interact with each other. This will aid in achieving positive outcomes from mitigation measures (Morrison-Saunders, Baker and Arts, 2003). The effective management and implementation of mitigation measures depends on the degree of coordination and relationship between the actors involved in the EIA system (Brown et al., 1991; Ortolano et al., 1987). This is more than just a good EIA report and Environmental Management Plan. Essentially, the interaction and mutual cooperation among these driving forces would help in the proper implementation of mitigation measures in Bangladesh (Figure 8.3).

8.7 OVERALL STATUS OF EIA SYSTEM IN BANGLADESH

The current situation (findings in Chapters 3–7) reveals that an effective EIA system in Bangladesh is on the right track. Indeed, the current situation of the EIA system is better than it had been in the past. There has been legislation for EIAs since 1995. The DOE is well mandated by the EIA legislation (ECA) to exercise its power in implementing the EIA. The capacity of the DOE is better than it was 15 years ago.

Despite this progress, the EIA system in Bangladesh is still far from good practice: the EIA legislation is not comprehensive, the DOE is not well funded, and there is a shortage of staff. In addition, the function of the DOE is not decentralized and the EISs are still deficient in quality information. Moreover, mitigation measures are poorly implemented and, therefore, predicted impacts of a project are not addressed adequately.

8.8 PROPOSED FRAMEWORK OF AN EFFECTIVE EIA SYSTEM

In this study, a conceptual framework integrating institutional arrangements, quality of EISs, and the implementation of mitigation measures was developed and used to evaluate the EIA system in Bangladesh. Using this framework, this study has successfully explicated key features of an EIA system in Bangladesh. Figure 8.4 shows the proposed effective EIA system in Bangladesh. The figure also shows that an effective EIA system in Bangladesh requires appropriate institutional arrangements, assuming that the provisions of the EIA process are clearly and adequately prescribed by the legislations (ECA and ECR). The DOE needs to be adequately staffed as well as budgeted so that it can regularly and effectively supervise EIA practice. The functions of the DOE need to be decentralized in order to enhance its scope and capacity to monitor, supervise, and support the proponents. Decentralization of the DOE's function is also necessary to support community participation in the EIA process.

The procedural requirements of the EIA process (such as scoping, collection of base line information, the identification and assessment of impacts using appropriate tools and techniques, the analysis of alternatives, and the preparation of an Environmental Management Plan) should be effectively addressed. This results in the EISs containing good quality information for decision makers. Similarly, the EIA system requires expert EIA consultants, the preparation of TORs with the vetting of the DOE, the expertise of the DOE in reviewing EISs, and adequate time to study EIAs in order to improve the quality of EISs. Furthermore, changes in the attitude of proponents along with the consultants when preparing EISs is also required in making the EIA system effective in Bangladesh.

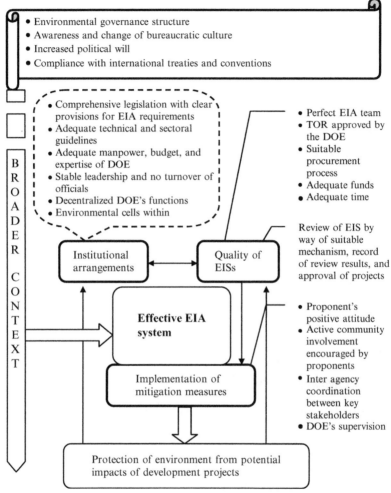

Figure 8.4 A framework of effective EIA system. *Developed by the researchers.*

Community participation is necessary not only to identify potential impacts of a project but also to implement the mitigation measures successfully and, thereby, make the EIA system effective. Continuous monitoring and feedback by the proponents is required to implement the mitigation measures successfully and, thereby, address the environmental and social impacts. Moreover, an effective interagency coordination between the donor agencies, DOE, proponents, and other entities, such as the Planning Commission, is important during the whole life cycle of the project.

As well as all of the above, a favorable broader context is necessary to make the EIA system effective in Bangladesh. In particular, an environmental governance system is not well established yet in Bangladesh. Such a system would help to make the people aware of the environmental issues and their rights. These broader contextual factors, while taking time to develop, are essential in making the EIA system effective.

8.9 CONCLUSION TO THE RESEARCH PROBLEM: AN EFFECTIVE EIA SYSTEM IN BANGLADESH

The findings that emerged from this investigation (Chapters 3–7) have provided insights into the current status of the EIA system in Bangladesh. To conclude, the findings of this study indicate that the EIA system in Bangladesh is on the right track. However, the EIA system is still far from fulfilling good practice requirements. There has been limited success in providing an environmental agency with adequate capability and in providing quality information. In addition, mitigation measures are not adequately implemented at the post-EIS stage of the EIA process. Therefore, the government of Bangladesh needs to take necessary actions to make the EIA system more effective. With this in mind, a number of recommendations are made below.

8.10 RECOMMENDATIONS

8.10.1 Improvement in Institutional Arrangements

1. The EIA legislation should be amended to include adequate and clear provisions of EIA requirements. The stages of the EIA process and other requirements, such as provision of EIAs for the extension of project and the review process of EIA reports, need to be defined by the current EIA legislations. The sectoral guidelines then may illustrate the methodological requirements of each stage of the EIA process and the responsibilities of stakeholders, and the procedure for the detailed review of EISs.

2. The government can take immediate action in enhancing the capacity of the DOE. The DOE should be equipped with adequately trained staff and ample budget, and stable leadership with a view to enforcing the proponents' efficient and practical compliance with EIA requirements. Setting up environmental units within the Planning Cell at different ministries may facilitate better interagency coordination.

3. Decentralization of the DOE's function and administration at district and Upa-Zilla (subdistrict) levels (two important administrative tiers of

central government) is an urgent need. This will not only enhance the DOE's capacity to closely monitor and supervise the implementation of mitigation measures of a project but also easily and actively facilitate the participation of community members in the process.

8.10.2 Improvement in the Quality of EISs

1. It is necessary for the DOE to establish an up-to-date environmental database as the source of baseline data and to make the information accessible by all EIA practitioners. There should be the practice of publishing EISs in the Bengali language in order to make the EISs easily accessible to all professions in the broader public in Bangladesh. At least, the nontechnical summary should be published in the Bengali language.

2. An attitudinal change favorable to good EIA practice on the part of the proponent is necessary. This is possible only when they are well aware of the environmental consequences of a development project and the ultimate benefits of the EIA application. Training and education program by the DOE in order to develop awareness among the proponents are an urgent need.

3. In order to control the bias of EIA consultants in the preparation of EISs, ethical codes of conduct should be initiated by the government. The introduction of an accreditation system to maintain the quality of EIA consultants may also improve the quality of EISs.

8.10.3 Improvement in the Implementation of Mitigation Measures

1. Active participation of the local community can make the implementation and monitoring program effective. Therefore, the proponents should involve the members of affected community in the implementation of mitigation measures.

2. Donor agencies should control the release of funds based on environmental performance of proponents. This will enforce the proponents to make the implementation of mitigation measures and monitoring activity effective. Close communication between donor agencies and proponents is required.

3. Systematic interagency cooperation and coordination among the core stakeholders (the DOE, donor agencies, the community, and the proponent) is essential for improving EIA practice in Bangladesh. In addition, regular meetings of the National Environmental Council undertaken by

the initiative of MoEF may facilitate interagency cooperation and coordination in order to implement the mitigation measures and other management activities.

8.11 CHAPTER SUMMARY

This chapter has provided an overview of the EIA system in Bangladesh based on the study's findings and their analyses. This chapter provided an insight into the EIA system in Bangladesh together with the institutional aspects, the quality of EISs, and the implementation of mitigation measures. The findings have been substantiated with reference to other studies. The discussion also focused on the shortcomings of the EIA system in Bangladesh and explained the possible reasons for these shortcomings. The chapter finally made recommendations for improvement to address the deficiencies of the EIA system.

REFERENCES

Abracosa R, Ortolano L. Environmental impact assessment in the Philippines: 1977–1985. Environ Impact Asses Rev 1987;7(4):293–310.

ADB. The Philippines environmental impact statement system: framework, implementation, performance and challenge, Manila, the Philippines; 2007.

Ahmad B, Wood C. A comparative evaluation of the EIA systems in Egypt, Turkey and Tunisia. Environ Impact Assess Rev 2002;22:213–34.

Badr E, Cashmore M, Cobb D. The consideration of impacts upon the aquatic environment in environmental impact statements in England and Wales. J Environ Assess Policy Manage 2004;6(1):19–49.

Barker A, Wood C. An evaluation of EIA system performance in eight EU countries. Environ Impact Assess Rev 1999;19:387–404.

Briffett C. Environmental impact assessment in East Asia. In: Petts J, editor. Handbook of environmental impact assessment: environmental impact assessment in practice - impact and limitations, vol. 2. Oxford: Blackwell; 1999. p. 143–67.

Brown AL, Hindmarsh RA, McDonald GT. Environmental assessment process and issues in the Pacific Basin-Southeast Asia Region. Environ Impact Assess Rev 1991;11:143–56.

Dipper B, Jones C, Wood C. Monitoring and post-auditing in environmental impact assessment: a review. J Environ Plan Manage 1998;41:731–47.

Doberstein B. Environmental capacity-building in a transitional economy: the emergence of the EIA capacity in Vietnam. Impact Assess Project Appraisal 2003;21(1):25–42.

Fuller K. Quality and quality control in environmental impact assessment. In: Petts J, editor. Handbook of environmental impact assessment, vol. 2. Oxford: Blackwell Science Ltd.; 1999. p. 55–84.

Glasson J, Salvador NNB. EIA in Brazil: a procedure-practice gap. A comparative study with reference to the European Union, and especially the UK. Environ Impact Assess Rev 2000;20:191–225.

Glasson J, Therivel R, Weston J, Wilson E, Frost R. EIA-learning from experience: changes in the quality of environmental impact statements for UK Planning Projects. J Environ Plan Manage 1997;40(4):451–64.

Jalava K, Pasanen S, Saalasti M, Kuitunen M. The quality of environmental impact assessment: Finnish environmental impact statements and the opinions of EIA professionals. Impact Assess Project Appraisal 2010;28(1):15–27.

Lee N, George C, editors. Environmental assessment in developing and transitional countries: principals, methods and practice. New York: John Wiley and Sons, Ltd; 2000.

Lohani BN, Evans JW, Everitt RR, Ludwig H, Carpenter RA, Tu S. Environmental impact assessment for developing countries in Asia. , vol. 1. Manila: Asian Development Bank; 1997. p. 1–356.

Memon PA. Devolution of environmental regulation: environmental impact assessment in Malaysia. Impact Assess Project Appraisal 2000;18(4):283–93.

Modak P, Biswas AK. Conducting environmental impact assessment for developing countries. New York: United Nations University Press; 1999.

Momtaz S. Environmental impact assessment in Bangladesh: a critical review. Environ Impact Assess Rev 2002;22:163–79.

Momtaz S. Public participation and community involvement in environmental and social impact assessment in developing countries. Int J Environ Cult Econ Soc Sustain 2006;2(4):89–97.

Morrison-Saunders A, Annandale D, Cappelluti J. Practitioner perspective on what influences EIA quality. Impact Assess Project Appraisal 2001;19(4):321–5.

Morrison-Saunders A, Baker J, Arts J. Lessons from practice: towards successful follow-up. Impact Assess Project Appraisal 2003;21(1):43–56.

Ortolano L. Controls on project proponents and environmental impact assessment effectiveness. Environ Professional 1993;15:352–63.

Ortolano L, May CL. Appraising effects of mitigation measures: the Grand Coulee Dam's impacts on fisheries. In: Morrison-Saunders A, Arts J, editors. Assessing impact: handbook of EIA and SEA follow-up. London: Earthscan; 2004.

Ortolano L, Jenkins B, Abracosa RP. Speculations on when and why EIA is effective. Environ Impact Assess Rev 1987;7:285–92.

Sadler B. The evaluation of assessment: post-EIS research and process development. In: Wathern P, editor. Environmental impact assessment: theory and practice. Boston: Unwin Hyman; 1988. p. 129–42.

Sadler B. International study of the effectiveness of environmental assessment: final report, Canadian Environmental Assessment Agency/International Association for Impact Assessment; 1996.

Sandham LA, Pretorius HM. A review of EIA report quality in the North West province of South Africa. Environ Impact Assess Rev 2008;28(4–5):229–40.

Vanclay F, Esteves AM, editors. New directions in social impact assessment. Cheltenham, UK: Edward Elgar; 2011.

Wang Y, Morgan RK, Cashmore M. Environmental impact assessment of projects in the People's Republic of China: new law, old problem. Environ Impact Assess Rev 2003;23(5):543–79.

Analysis of Alternative Alternative analysis is a part of EIA process. This involves the analysis of alternative sites, process of construction of an infrastructural project, or production process for an industrial project or even alternative mitigation measures to avoid the potential social and environmental impacts of a project.

Auditing Auditing is a term borrowed from accounting to describe a systematic process of examining, documenting, and verifying that EIA procedures and outcomes correspond to objectives and requirements. This can be implementation audits, impact audit, and compliance audits. Concept which refers to objective examination—a comparison of observations with predetermined criteria. Audits are dependent on monitoring data.

Baseline Data Data those are collected to identify and assess potential impacts of a project. Baseline data can be collected from either primary or secondary sources.

Checklist Method Checklists are standard lists of the types of impacts associated with a particular type of project. Checklist methods are primarily for organizing information or ensuring that no potential impact is overlooked.

Community Participation Community participation is a vital element of EIA process. The aim of the community participation is to get more information about potential environmental impacts of the projects, identify potential mitigation measures, and improve the EIA practice. Ideally, community should be involved at each stage of EIA process during the identification of impacts and implementation of mitigation measures.

Cumulative Impact Cumulative impacts are results from the interaction of a number of projects concentrated in a particular geographic area at a time or over a period of time. This impact for an individual project may be minor, but it can be found significant when assessed for a number of projects collectively in a geographic or temporal perspective.

EIA Guideline A guideline is a document that contains the EIA process in detail. An EIA guideline either technical or general involves the steps of the process in details, for example, the process of EIA community participation or the EIA process as whole delineating each of the steps.

EIA process The process that involves a number of stages such as screening, scoping, identification, and assessment impacts, preparation of mitigation measures, and monitoring.

EIA system EIA system involves both institutional arrangements and practice of EIA. EIA system = institutional arrangements + EIA practice (preparation of EIS + implementation of mitigation measures).

Environmental Clearance Certificate After the submission of EIA report, the environmental agency issues a certificate called Environmental Clearance Certificate. It contains a set of conditions prescribed by the environmental agency mainly related to the mitigation of potential impacts of the project.

Environmental Impact Assessment An EIA is an environmental management tool that helps to identify and predict potential environmental impacts of a project.

Environmental Impact Statement An environmental impact statement is the output of Environmental Impact Assessment that contains information on potential environmental impacts of a project. Environmental impact statement is also called environmental impact assessment report.

189

Environmental Management Plan An Environmental Management Plan is prepared based on the potential significance of impacts of a project. The EMP usually involves mitigation measures to mitigate potential significant impacts. This also includes the strategies of the implementation of mitigation measures including institutional arrangements and monitoring program.

Evaluation of Impact Evaluation of impact means to evaluate the significance of an impact. An impact is significant when it crosses the environmental quality standard or causes irreversible social and environmental damage.

Holistic model A holistic model represents a holistic view of an EIA system involving institutional arrangements, practice of EIA at both predecision stage and postdecision stage of EIA process and influence of EIA in broader decision making in order to change norms and values of stakeholders. This is an approach to review the EIA system that represents a comprehensive idea about the system.

Initial Environmental Examination A rapid process to identify environmental impacts of projects. This is equivalent to a details scoping undertaken in some countries, for example, Canada. Based on IEE reports, decision is taken whether full EIA is required or not.

Mitigation Measures Mitigation measures are identified and prepared based on the significant potential environmental impacts. Usually, mitigation measures are taken to avoid, reduce, and compensate for adverse impacts of an action on the environment.

Monitoring Monitoring is an activity designed to identify the nature and causes of change. In the context of environmental assessment, it is a data collection activity undertaken to provide specific information on the characteristics and functioning of environmental and social variables in space and time.

NEPA (National Environmental Policy Act) Statute passed by the US Congress in 1969 establishing the basic environmental policy for protection of the environment. The Act contains action-forcing procedures that must be followed up by federal agencies to ensure federal decision makers to take environmental factors into account before making a final decision regarding a proposed action.

Nontechnical Summary A nontechnical summary is a part of Environmental Impact Statement that contain the whole EIS in brief. The aim of the NTS is to provide the basic information on the potential impacts of a project and recommendations to address the impacts.

Prediction of Impact Impact prediction in EIA typically involves the use of model to either mathematically or conceptually represent the biophysical or socioeconomic environment. When an impact is identified, the next step is to predict its magnitude. Various models and expert judgement are used to predict the impacts of a project.

Project Proponent A project proponent is one who invests money and is responsible for the planning and implementation of the project.

Quality of EIS The quality of EIA reports in terms of adequacy and clarity of information the EIA reports contain.

Review of EIS Review of EIS is a process where the EIS is reviewed by a competent authority to judge whether the requirements of the EIA process are adequately addressed during EIA study.

Scoping Scoping is a process within the EIA process that identifies the boundary of impacts, strategies of data collection, and community participation and prepares a Terms of Reference. Scoping helps take preparation to conduct an EIA in details.

Screening Screening is the selection of projects which needs Environmental Impact Assessment either at initial scale or at full scale. In some jurisdiction, screening is done project by project depending on the discretion of the competent agency. In some jurisdiction, projects are listed required EIA or no EIA which is called normative screening.

SEA This is a tool to assess the potential environmental and social impacts of policies, programs, and plans. This can be called a program or policy-level EIA.

Significant Impact An impact is considered significant when it crosses the environmental standards or limits of a particular jurisdiction. The significance of an impact however depends on the value of the community.

Social Impact Assessment Social Impact Assessment (SIA) includes the processes of analyzing, monitoring, and managing the intended and unintended social consequences, both positive and negative, of planned interventions.

Terms of Reference A document that involves a set of instruction given to someone. It is a mutually agreed document with clear objective and tasks to be done of an activity.

Uncertainty In EIA process, uncertainty occurs when the baseline data are not correct or the process of prediction is not suitable to the context. Because of uncertainty, adaptive management approach can be used to mitigate the unpredictive or under predictive.

INDEX

Note: Page numbers followed by *b* indicate boxes, *f* indicate figures and *t* indicate tables.

Printed and bound by CPI Group (UK) Ltd, Croydon, CR0 4YY

03/10/2024

01040422-0002